Distributed Network Data

Alasdair Allan and Kipp Bradford

Beijing · Cambridge · Farnham · Köln · Sebastopol · Tokyo

Distributed Network Data

by Alasdair Allan and Kipp Bradford

Copyright © 2013 Alasdair Allan and Kipp Bradford. All rights reserved.

Printed in the United States of America.

Published by O'Reilly Media, Inc., 1005 Gravenstein Highway North, Sebastopol, CA 95472.

O'Reilly books may be purchased for educational, business, or sales promotional use. Online editions are also available for most titles (*http://my.safaribooksonline.com*). For more information, contact our corporate/institutional sales department: 800-998-9938 or *corporate@oreilly.com*.

Editor: Julie Steele
Production Editor: Kristen Borg
Proofreader: O'Reilly Production Services

Cover Designer: Randy Comer
Interior Designer: David Futato
Illustrator: Rebecca Demarest

March 2013: First Edition

Revision History for the First Edition:

2013-03-08: First release

See *http://oreilly.com/catalog/errata.csp?isbn=9781449360269* for release details.

Nutshell Handbook, the Nutshell Handbook logo, and the O'Reilly logo are registered trademarks of O'Reilly Media, Inc. *Distributed Network Data*, the image of a guillemot, and related trade dress are trademarks of O'Reilly Media, Inc.

Many of the designations used by manufacturers and sellers to distinguish their products are claimed as trademarks. Where those designations appear in this book, and O'Reilly Media, Inc., was aware of a trademark claim, the designations have been printed in caps or initial caps.

While every precaution has been taken in the preparation of this book, the publisher and authors assume no responsibility for errors or omissions, or for damages resulting from the use of the information contained herein.

ISBN: 978-1-449-36026-9

[LSI]

Table of Contents

Preface . . . . **vii**

1. Introduction to Arduino **1**
    Saying Hello in the Old World   1
    The Arduino   2
        The Board   2
        Powering the Board   3
        Input and Output   4
        Communicating with the Board   4
        Installing the Software   4
        Connecting to the Board   5
    Blinking an LED   7
        Uploading the Sketch   10
    Making a Serial Connection   12
    Summary   14

2. Getting Started **15**
    The Breadboard   15
    The Sensor   17
        The DHT-22   17
    Wiring the Breadboard   18
    Writing the Software   21
        The DHT Library   22
        Arduino Sketch   22
    Running the Software   24
    Summary   25

3. Adding Another Sensor **27**
    The Sensor   27

PIR Sensor	27
Wiring the Breadboard	28
Modifying the Software	30
Running the Software	32
Summary	33

4. Finishing the Breadboard... 35
The Sensor	35
Electret Microphone	35
Wiring the Breadboard	36
Modifying the Software	38
Running the Software	41
Adding Some LEDs	42
Modifying the Software	44
Running the Software	47
Making the Output Machine-Readable	48
Communicating with Python	50
Summary	52

5. Moving from Breadboard to Prototype.. 53
The Prototype	53
Fritzing	54
Installing the Software	55
Building a Fritzing Circuit	56
Cleaning up the Fritzing Diagram	62
Dropping Breadboard Crumbs	65
Making the Board Permanent	65
Solder	65
Soldering Irons	66
The Protoboard	67
Octopus Arms or Helping Hands	68
Let the Solder Flow	69
Summary	73

6. Simplifying the Design... 75
Arduino Proto Shields	75
Fritzing Revisited	77
Power and Ground	82
Cleaning up the Fritzing Diagram	88
Rules for Pin Power and Ground	89
Saving Power	90

    Summary   90

7. Building Point-to-Point XBee Networks... 91
    XBee Modules   92
        Series 1 or Series 2?   93
        Regular vs Pro?   94
        802.15.4 or Zigbee?   95
        Which Antenna?   95
    How to Configure an XBee Series 1 Radio   95
        Connecting the XBee to your Mac   96
        XBee Addressing   100
    Configuring Two XBee Radios   100
    Connecting an XBee to an Arduino   103
    Going Wireless with XBees   106
    Summary   107

8. Building Many-to-Point XBee Networks... 109
    Addressing for Multiple XBees   109
    Addressing the Arduino   110
    Individual Call and Response   112
    Switching to Mesh Networks   116
    Summary   117

9. Visualizing with Processing... 119
    Processing   119
        Installing the Software   120
    Reading Data From a File   120
    Reading Data Directly From the Serial Port   128
        Plotting Temperature in Real Time   129
    Summary   132

10. Visualizing with LabVIEW... 133
    LabVIEW   133
        Installing the Software   134
    Simple LabVIEW with Arduino   145
    Graphing the Data   148
    Summary   150

11. Going Further.. 151
    Arduino   151
    XBee Networking   151
    Fritzing   151

EAGLE	152
Processing	152
ProcessingJS	152
LabVIEW	152
Data Visualization	153

Preface

The gap between having an idea and being able to build a hardware solution is narrowing. The ability to prototype, build, and deploy simple sensor platforms is rapidly leading to an exponential growth in the amount of data available.

Over the next few years, day-to-day computing will become increasingly invisible, dissipating out into the environment. This is already starting to happen, without you noticing, as the physical interfaces to the new smart devices look almost identical to their dumb counterparts.

You will soon begin to move in a sea of data: your movements tracked and your environment measured and adjusted to your preferences, all without your intervention.

At the O'Reilly Strata Conference on data science in Fall 2012 in New York, we gave the attendees a taste of the super-connected world that's ahead of all of us. By instrumenting the conference environment with basic off-the-shelf sensors and mesh networking, we observed and reported, and gave the attendees a taste of their lives in a more measured and quantified world.

This book will allow you to do likewise.

Bill of Materials

Before we get started building our sensor mote we'll need to gather all the components we'll need together in one place. In the hardware business this is generally called the "bill of materials." Initially we'll make use of the following components:

- Arduino Uno
- Breadboard
- DHT-22
- PIR sensor

- Adafruit Electret Microphone
- 220Ω Resistor (×3)
- 10kΩ Resistor
- LED (×3)
- Jumper wire

Later in the book we'll expand our sensor mote to communicate over XBee (802.15.4) networking and if you want to follow along you'll need to add the following additional components:

- XBee USB Explorer (*https://www.sparkfun.com/products/8687*), $24.95
- Arduino Wireless SD Shield (*http://arduino.cc/en/Main/ArduinoWirelessShield*), $19.95
- XBee Series 1 Radio (*http://www.adafruit.com/products/128*) (×2), $22.95

Who Should Read this Book?

This book provides an introduction to the topic of how to build and deploy a distributed sensor network. As part of that, we'll make extensive use of the Arduino (*http://arduino.cc/*) open-source electronics prototyping platform. This is hardware hacking for data scientists. If you're a designer familiar with data visualization or a programmer with in interest in data, and want to learn about how to build simple sensor networks to gather data about your environment, then this book is for you.

What You Should Already Know?

This book is intended for data scientists who want to learn how to work with external hardware. It assumes some basic computing and programming knowledge, but no real expert knowledge is assumed.

Little or no familiarity with the Arduino platform is expected. However, if you are totally unfamiliar with the Arduino platform, you might want to take a look at *Getting Started with Arduino* by Massimo Banzi (O'Reilly).

What Will You Learn?

This book will guide you through building a distributed sensor network and gathering data, and will show you how to do some simple real-time analysis and visualization. It will walk you through your first hardware prototypes, show you how to improve them, and teach you how to build a network of sensors and begin taking data.

What's In This Book?

Chapter 1, Introduction to Arduino
> This chapter is intended for programmers new to Arduino. It will introduce you to the platform and walk you through the hardware equivalent of "Hello World": the blinking LED. We'll also discuss how to use the serial connection between the Arduino and your development machine.

Chapter 2, Getting Started
> This chapter will walk you through wiring up a breadboard to prototype a circuit, and using the Arduino to read values from a sensor.

Chapter 3, Adding Another Sensor
> This chapter provides a hands-on tutorial to adding a second sensor to the breadboard design—this one an infared motion detector (PIR).

Chapter 4, Finishing the Breadboard
> This chapter is a guide to adding yet another sensor to your mote – a microphone. It also walks you through added diagnostic LEDs and modifying your Arduino sketch to output CSV through the serial console.

Chapter 5, Moving from Breadboard to Prototype
> This chapter will explain how to use Fritzing, a program that lets us convert our circuit design into a direct graphical representation on the computer in preparation for building an actual circuit board. It also introduces theArduino prototyping shield and shows you how to turn your sketch into a professional circuit board.

Chapter 6, Simplifying the Design
> After explaining the fundamental concepts of power and ground, this chapter takes a look at Arduino pins and how to use them for power, discussing the limitations of this approach as well as some particular use cases.

Chapter 7, Building Point-to-Point XBee Networks
> This chapter discusses eliminating the USB cable and replacing it with a wireless connection to the Internet using XBees.

Chapter 8, Building Many-to-Point XBee Networks
> This chapter provides a basic overview of how to handle multiple sensor platforms and request data using a single master multiple slave configuration, as well as a multi-master system, using zigbee protocols.

Chapter 9, Visualizing with Processing
> This chapter will show you how to take the data your sensor network has collected and begin to visualize it using Processing.

Chapter 10, Visualizing with LabVIEW
> This chapter gives you some more data visualization options with an introduction to LabVIEW.

Chapter 11, Going Further
> This chapter provides a collection of pointers to more advanced material on the topics we covered in the book, and material covering some of those topics that we didn't manage to talk about in this book.

Conventions Used in This Book

The following typographical conventions are used in this book:

Italic
> Indicates new terms, URLs, email addresses, filenames, and file extensions.

`Constant width`
> Used for program listings, as well as within paragraphs to refer to program elements such as variable or function names, databases, data types, environment variables, statements, and keywords.

`Constant width bold`
> Shows commands or other text that should be typed literally by the user.

`Constant width italic`
> Shows text that should be replaced with user-supplied values or by values determined by context.

 This icon signifies a tip, suggestion, or general note.

 This icon signifies a warning or caution.

Using Code Examples

This book is here to help you get your job done. In general, if this book includes code examples, you may use the code in your programs and documentation. You do not need to contact us for permission unless you're reproducing a significant portion of the code. For example, writing a program that uses several chunks of code from this book does not require permission. Selling or distributing a CD-ROM of examples from O'Reilly books does require permission. Answering a question by citing this book and quoting example code does not require permission. Incorporating a significant amount of example code from this book into your product's documentation does require permission.

We appreciate, but do not require, attribution. An attribution usually includes the title, author, publisher, and ISBN. For example: "*Distributed Network Data* by Alasdair Allan and Kipp Bradford (O'Reilly). Copyright 2013 Alasdair Allan and Kipp Bradford, 978-1-449-36026-9."

If you feel your use of code examples falls outside fair use or the permission given above, feel free to contact us at *permissions@oreilly.com*.

Safari® Books Online

Safari Books Online (*www.safaribooksonline.com*) is an on-demand digital library that delivers expert content in both book and video form from the world's leading authors in technology and business.

Technology professionals, software developers, web designers, and business and creative professionals use Safari Books Online as their primary resource for research, problem solving, learning, and certification training.

Safari Books Online offers a range of product mixes and pricing programs for organizations, government agencies, and individuals. Subscribers have access to thousands of books, training videos, and prepublication manuscripts in one fully searchable database from publishers like O'Reilly Media, Prentice Hall Professional, Addison-Wesley Professional, Microsoft Press, Sams, Que, Peachpit Press, Focal Press, Cisco Press, John Wiley & Sons, Syngress, Morgan Kaufmann, IBM Redbooks, Packt, Adobe Press, FT Press, Apress, Manning, New Riders, McGraw-Hill, Jones & Bartlett, Course Technology, and dozens more. For more information about Safari Books Online, please visit us online.

How to Contact Us

Please address comments and questions concerning this book to the publisher:

> O'Reilly Media, Inc.
> 1005 Gravenstein Highway North
> Sebastopol, CA 95472
> 800-998-9938 (in the United States or Canada)
> 707-829-0515 (international or local)
> 707-829-0104 (fax)

We have a web page for this book, where we list errata, examples, and any additional information. You can access this page at *http:/oreil.ly/distributed-network-data*.

To comment or ask technical questions about this book, send email to *bookquestions@oreilly.com*.

For more information about our books, courses, conferences, and news, see our website at *http://www.oreilly.com*.

Find us on Facebook: *http://facebook.com/oreilly*

Follow us on Twitter: *http://twitter.com/oreillymedia*

Watch us on YouTube: *http://www.youtube.com/oreillymedia*

Acknowledgments by Alasdair Allan

Everyone has one book in them, however this isn't mine. This isn't my first book, but it's my first with a coauthor, and that's a very different experience. Books do not write themselves, and I'd like to thank my coauthor, Kipp Bradford, and my editors at O'Reilly, Julie Steele and Brian Jepson, for holding my hand through the process.

I very much want to thank my wife Gemma Hobson for her continued support and encouragement, and for letting me fly to the States for the week before Christmas to camp out in my coauthor's living room without complaint. Those small (and sometimes larger) sacrifices an author's spouse routinely has to make don't get any less inconvenient the second, or third, or the nth time around. I'm not sure why she lets me write, perhaps because I claimed to enjoy it so much. Thank you Gemma. Finally to my son Alex, who is just now learning to read, and won't be reading these words for some time to come. Thank you.

Acknowledgments by Kipp Bradford

This is my first book. For all the projects I've done and stories I'd like to tell, this is the first time I've actually sat down to write something. Hopefully I have many more books ahead of me. I sincerely want to thank my coauthor, Alasdair Allan, for trusting me enough to share this adventure with him and showing me how to write. I still have a lot to learn, and I'm depending on you to continue being a wonderful friend and mentor! I also must thank my editor, Julie Steele, for being so incredible and supportive throughout the process of liberating the words from my brain and fingers, and turning those words into something that people will want to read. I do need to thank Brian Jepson for introducing me to all the wonderful people at O'Reilly and encouraging my technical writing tendencies.

Finally, thank you to my dear friends and family who kept me fed, loved, and relaxed as I worked. This book is dedicated to you.

CHAPTER 1
Introduction to Arduino

The ubiquitous "Hello World" application that every programmer starts off with when faced with a new language has a hardware equivalent: blinking an LED on and off. It doesn't sound too hard. But ten years ago, building even simple hardware was a lot harder than it is now.

Saying Hello in the Old World

Typically, a microprocessor used to arrive on a proprietary development board, along with a compiler on a CD ROM. Loading the compiler onto your computer took several hours. Then you had to locate the correct drivers to tell your computer how to talk to the evaluation hardware. Then you had to install the drivers and—if everything worked—by the end of the first day your compiler might just be talking to the evaluation board with your microprocessor.

The next day you needed to download the manufacturer's sample applications, which would be written in machine code. From there, you probably had to spend another day configuring you compiler and addressing linking issues before getting the code working on the evaluation hardware. If everything went smoothly, it would take only a day or two to get to a blinking LED. If you weren't as lucky, it might take a week, even for someone familiar with hardware.

Then, if you wanted to change the blink rate of your LED, you have to do a lot of reading about the architecture of your chip—one of hundreds of different processor all with subtly different architectures—and figure out how registers and timer interrupts worked for your architecture. This whole process took days, if not weeks, and in the mid-nineties a reasonably brief walkthrough of the process would take several hundred pages to explain it properly. Even ten years after that, it was still probably the best part of seventy pages to reach a working application. We're going to manage it over the course of the next few pages.

The Arduino

In the last several years, all this has changed. Every so often, a piece of technology can become a lever that lets people move the world, just a little bit. The Arduino is one of those levers.

It started off as a project to give artists access to embedded microprocessors for interaction design projects, but it may well end up in a museum as one of the building blocks of the modern world. It allows rapid, cheap prototyping for embedded systems. It turns what used to be fairly tough hardware problems into simpler software problems (and we know that once problems are in software, they become almost exponentially easier).

The Arduino—and the open hardware movement that has grown up with it, and at least to a certain extent around it—is enabling a generation of high-tech tinkerers both to break the seals on proprietary technology and to prototype new ideas with fairly minimal hardware knowledge. This Maker renaissance has led to an interesting growth in innovation. People aren't just having ideas, they're doing something with them.

The Board

The current revision of the Arduino board is known as the Arduino Uno (see Figure 1-1). This board is based on the ATmega328 microcontroller.

Figure 1-1. The Arduino Uno board with the ATmega328 microcontroller

It has fourteen digital input/output pins, six of which can be used as PWM outputs, along with six more analog input pins (see Table 1-1).

Table 1-1. Technical specification of the Arduino Uno board

Arduino Uno	
Microcontroller	ATmega328
Operating Voltage	5 V
Input Voltage (recommended)	7 – 12 V
Input Voltage (limits)	6 – 20 V
Digital I/O Pins	14 (6 provide PWM)
Analog Input Pins	6
DC Current per I/O Pin	40 mA
DC Current for 3.3V Pin	50 mA
Flash Memory	32 KB
SRAM	2 KB
EEPROM	1 KB
Clock Speed	16 MHz

The Arduino platform has everything needed to support the on-board microcontroller. It is attached directly to your Mac for programming using a USB connection, and it can be powered via the same USB connection, or with an external power supply if you want to detach the board from your Mac once you've programmed it.

To keep things short, this tutorial will assume you're using a recent Arduino board such as the Arduino Uno, Duemilanove, or Diecimila. However, if you're working with an older board or one of the many Arduino-compatible clones, it's likely that little needs to be changed. See this Arduino site (*http://arduino.cc/en/Main/Hardware*) for links to the Getting Started guides for various other boards.

Powering the Board

The Arduino Uno can be powered via the USB connection, or with an external power supply. Unlike previous generations of the Arduino, the power source is selected automatically. If you're using an earlier model, you will have to manually change between USB and external power sources using a jumper on the board itself; this jumper is usually located between the USB and power jacks.

The board can operate on an external supply of 6 to 20 volts. However, if supplied with less than 7V, the 5V pin may supply less than nominal voltage and the board may become unstable. If using more than 12V, the voltage regulator may overheat and damage the board. The recommended range is therefore between 7 to 12 volts.

Input and Output

Each of the 14 pins on the Uno can be used as an input or output. They operate at 5V, with each pin having an internal pull-up resistor (disconnected by default) of 20-50 kOhms. The maximum current that a pin can provide is 40mA.

Some pins have specialized functions. Perhaps the most important of these to us for the purposes of this book are pins 0 and 1. These pins can be used to receive (RX) and transmit (TX) TTL serial data. These pins are connected to the corresponding pins of the ATmega8U2 USB-to-TTL Serial chip.

Communicating with the Board

The ATmega328 provides UART TTL serial communication at 5V, which is available on digital pins 0 (RX) and 1 (TX). The Arduino Uno has an ATmega8U2 chip on-board that redirects this serial communication over USB, allowing the Arduino to appear as a virtual serial port to software on your Mac. If you're using an older board, such as the Duemilanove or Diecimila, these boards make use of the FTDI USB-to-serial driver chip to accomplish the same task, although unlike with the newer Uno board, you will be required to install a driver so that your Mac can see the board correctly.

Installing the Software

Download the latest version of the development environment from the Arduino.cc website (*http://arduino.cc/en/Main/Hardware*).

The latest version of the Arduino development environment is Arduino 1.0.3, at the time of writing.

Like most Mac applications,[1] the development environment comes as a disk image (*.dmg* file) that should mount automatically after you have downloaded it. If it doesn't, double-click on it to open it manually. After it is open, just click-and-drag the *Arduino.app* application into your */Applications* folder. If you're using an older board that requires that you install the FTDI USB-to-Serial drivers, you should also double-click on the *FTDIUSBSerialDriver.mpkg* file included in the disk image to install the necessary drivers.

---

1. Full instructions for installation on Windows (*http://arduino.cc/en/Guide/Windows*) and on Linux (*http://playground.arduino.cc/Learning/Linux*) are available online.

Once you've installed the development environment, eject the disk image by dragging it to the trashcan and double-click on the *Arduino.app* application icon to start the IDE. You should see something that looks a lot like Figure 1-2.

Figure 1-2. The development environment

Connecting to the Board

Connect the Arduino to your Mac with an appropriate USB cable. In the case of an Arduino Uno, Duemilanove, or Diecimila, you'll need a USB-A (Mac) to USB-B (Arduino) cable: the same sort needed for most USB printers. The green power LED (labeled **PWR**) should go on, and if you're using the Arduino Uno, a dialog box will appear telling you that a new network interface has been detected. Just hit the Apply button. While the new interface will claim to be "Not Configured," if you inspect it in System Preferences, it is working correctly.

After connecting your board to your Mac, go to the Tools→Board menu item in the Arduino development environment and select your board from the list of supported board in the drop-down menu (see Figure 1-3).

Figure 1-3. Selecting the correct board type

Then go to the Tools→Serial Port menu and select the correct serial port for your board (Figure 1-4). On the Mac, the name will start with */dev/tty.usbmodem* for the Uno, or */dev/tty.usbserial* for older boards.

Figure 1-4. Choosing the correct serial port

If you're unsure which serial port corresponds to your board, you can always unplug and then re-plug the USB cable connecting your Mac to the Arduino board to see how the menu changes.

Blinking an LED

Now that we have our Arduino connected to our Mac, let's go ahead and walk through the hardware equivalent of "Hello World": the blinking LED.

An Arduino program is normally referred to as a *sketch*.

There are a number of example sketches included in the development environment. The one we're looking for can be found by selecting File→Examples→1.Basics→Blink. The example sketch will open in a new window (see Figure 1-5).

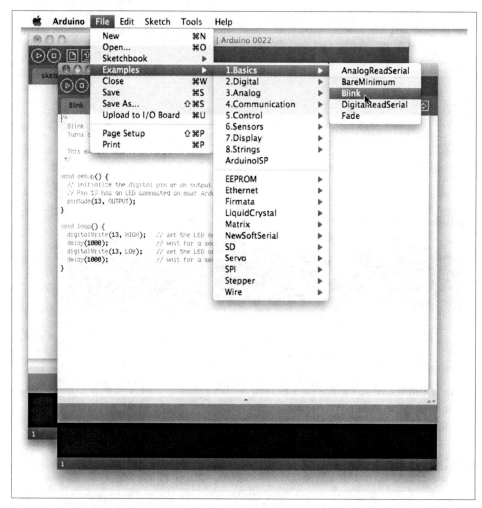

Figure 1-5. The Blink example

Every Arduino sketch consists of two parts: the setup and the loop. Every time the board is powered up, or the board's reset button is pushed, the `setup( )` routine is run. After that finishes, the board runs the `loop( )` routine. When that completes, and perhaps somewhat predictably, the `loop( )` is run again and again. Effectively, the contents of the `loop( )` sit inside an infinite `while` loop.

Before we go ahead to build and deploy this example to our Arduino, let's take a look at the code:

```
void setup() {
  pinMode(13, OUTPUT);❶
}

void loop() {
  digitalWrite(13, HIGH);❷
  delay(1000);
  digitalWrite(13, LOW);❷
  delay(1000);
}
```

❶ Arduino pins default to INPUT. However, here we set pin 13 to behave as an OUTPUT pin. In this state, the pin can provide up to 40 mA of current to other devices. This is enough current to brightly light up an LED, or run many sensors, but not enough current to run most relays, solenoids, or motors. See this Arduino site (*http://arduino.cc/en/Tutorial/DigitalPins*).

❷ If the pin has been configured for OUTPUT, its voltage will be set to the corresponding value: 5V for **HIGH**, 0V (ground) for **LOW**.

Effectively, then, this piece of code causes the voltage on pin 13 to be brought **HIGH** for a second (1,000 ms) and then **LOW** for a further second before the loop starts again, bringing the voltage **HIGH** once more.

As mentioned before, some of the digital pins on the Arduino board have special functions; pin 13 is one of these. On most boards (including the Uno), it has an LED and resistor attached to it that's soldered onto the board itself. Therefore, the effect of this sketch will be to turn the embedded LED on and off with a periodicity of 1 second.

While there is already a built-in LED on the board, this is going to look more impressive if we add a "real" LED as well. Any LED will do, but LEDs are directional components and must not be inserted backwards. Look carefully at the two legs of the LED; one should be shorter than the other. The shorter leg corresponds to the ground, while the longer leg is the positive.

Insert the short leg into the GND pin, and the longer leg into the pin 13, as shown in Figure 1-6.

Figure 1-6. An LED connected to pin 13. The short leg is inserted into the GND pin

Uploading the Sketch

The first step to getting the sketch ready for transfer over to the Arduino is to click on the Verify/Compile button (see Figure 1-7). This will compile your code, checking it for errors, and then translate you program into something that is compatible with the Arduino architecture. After a few seconds, you should see the message *"Done compiling."* in the Status Bar and something along the lines of *"Binary sketch size: 1018 bytes (of a 32256 byte maximum)"* in the Notification area.

Once you see that message, go ahead and click on the Upload button (see Figure 1-7 again). This will initiate the transfer of the compiled code to the board via the USB connection.

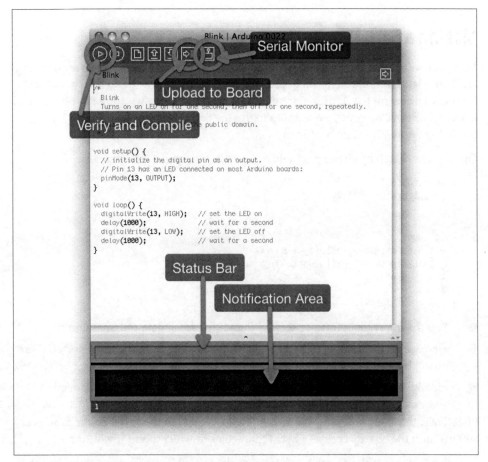

Figure 1-7. The Arduino development environment with various controls highlighted

Wait a few seconds—you should see the RX and TX LEDs on the board flashing as the data is transferred over the serial connection from your Mac to the board. If the upload is successful, the message "*Done uploading.*" will appear in the status bar.

A few seconds after the upload finishes, you should see the pin 13 LED on the board (labeled L on the PCB) start to blink (in orange) along with the LED we inserted into pin 13. One second on, one second off. If you see that: congratulations, you've just successfully gotten the hardware equivalent of "Hello World" to compile and run on the Arduino board.

Making a Serial Connection

Now that we've seen how a basic Arduino sketch works, let's move on to sending and receiving data from the Arduino board. We need to know how to do this, because this is how we'll control the Arduino or get sensor readings when we connect it to our iOS devices. For now, however, we're going to use the Serial Monitor (see Figure 1-7 again) inside the development environment.

Open a new sketch by clicking on File→New (⌘N) to open a new window.

```
void setup() {
  Serial.begin(9600); ❶
}

void loop() {
  while (Serial.available() <= 0) { ❷
    Serial.println("Hello world"); ❸
    delay(300);
  }
}
```

❶ Sets the data rate in bits per second (baud) for serial data transmission.

❷ Here we get the number of bytes available for reading from the serial port. If there are no bytes waiting in the buffer, we loop until some arrive.

❸ Finally, here inside the loop we print "Hello World" to the serial connection.

Effectively, this code will send the string "Hello World" to the serial connection every 300 ms until the board receives a byte (character), at which point it will stop.

Save the contents of the sketch to a file using the File→New (⌘S) menu item, and then click on the Verify button to compile your sketch—and, if all goes well, the Upload button to upload it to the board. You should see the RX and TX LEDs light up as the code is transferred. When you see the *Done uploading.* message in the status bar, click on the Serial Console button (see Figure 1-7 again) in the development environment to open the serial console.

Doing so will reset the Arduino, at which point you should see the phrase "Hello World" appear on the console every 300 ms (see Figure 1-8) accompanied by a flash of the TX LED on the Arduino board itself.

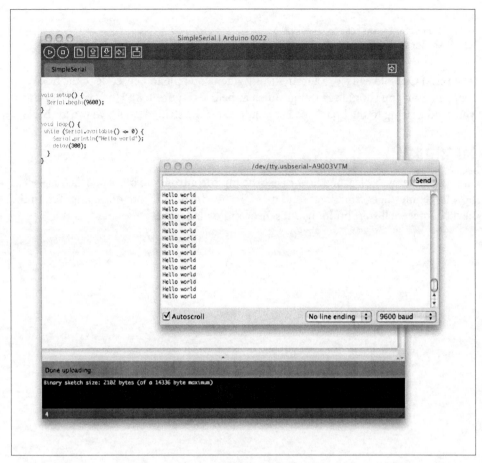

Figure 1-8. The Arduino says "Hello World" in the serial console

Entering a character or string in the text entry box and hitting the Send button will transmit those characters to the board, at which point the board should stop sending the string "Hello world" to the serial port.

Let's make that a bit more specific. Add the following bolded lines to your sketch and re-upload it to the Arduino:

```
void setup() {
  Serial.begin(9600);
}

void loop() {
  while (Serial.available() <= 0) {
    Serial.println("Hello world");
    delay(300);
  }
```

Making a Serial Connection | 13

```
    Serial.println("Goodbye world");
    while(1) { }
}
```

The Serial Console will close automatically when you upload new code to the board, so reopen it and you should see things much as before (Figure 1-8). However, this time, if you send a string to the board, you should receive the string "Goodbye world" back.

Summary

In this chapter we've learned how to use the microcontroller to blink an LED, as well as how to get messages from, and send messages to, the board from our computer. In the next chapter we'll start building out sensor mote platform.

CHAPTER 2
Getting Started

The Arduino board provides easy access to the digital and analog input and output pins of the microcontroller so that you can connect sensors and actuators to the processor.[1] However, it's not always possible to use those pins directly. Most sensors will need some additional supporting circuitry, and there are never enough power and ground pins to go around. What we need is some way to design and lay out our circuit —and perhaps more crucially, change our minds as we go along. The easiest way to do that is to use something called a *breadboard*.

The Breadboard

A solderless breadboard is an indispensable tool for rapidly and cheaply prototyping projects.

Breadboards consist of many tiny "holes" arranged on a 0.1-inch grid into which the leads of the component can be connected. Typically, as in Figure 2-1, the holes down each side of the breadboard are electrically connected lengthways down the board and are used for the positive and negative (i.e., ground or Earth) power supply. These are usually referred to as *rails*, and will commonly be labeled as such. The other holes in the board are connected horizontally across the board, usually (and again, as in Figure 2-1) with a gap down the middle. Each hole is connected to the many metal strips that are running underneath the board.

---

1. This is normally called *breaking out the pins*, and many surface mount components are made available on "breakout boards" to hobbyists.

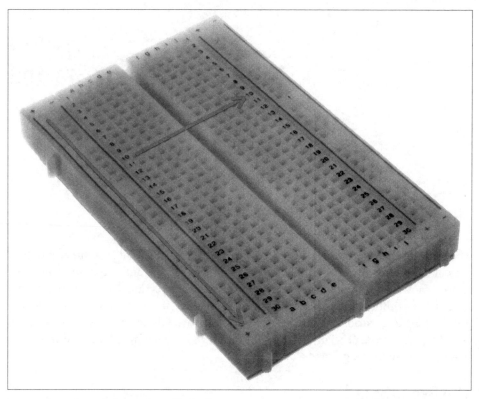

Figure 2-1. A typical mini breadboard

Putting the legs of a component in the same row, therefore, forms connections between different components of your circuit. Normally, you'll make use of short lengths of wire, commonly referred to as *jumper wires*, to connect rows of the board together.

 When using a breadboard, you must use single-core 0.6mm diameter wire. Stranded wire is not suitable because it will crumple when pushed into a hole, and it may damage the board if individual strands break off.

When using chips with many legs (*integrated circuits* or, more commonly, *ICs*), place them in the middle of the board so that half of the legs are on one side of the middle gap that runs down the board, and half are on the other side. Since the chip spans the gap in the middle of the board, the legs on one side of the chip will not be electrically connected to the legs on the other side of the chip.

The Sensor

We'll add several sensors over the course of the next few chapters, but for now let's start our project with a single sensor for measuring temperature and humidity.

The DHT-22

We're going to make use of the DHT-22 temperature and humidity sensor (see Figure 2-2).

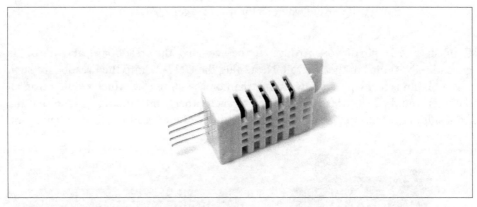

Figure 2-2. The DHT-22 temperature and humidity sensor

The DHT-22 is one of a range of basic, low cost, temperature/humidity sensors. It has 4 pins with 0.1 inch spacing, which means it can be inserted directly into our breadboard.

The sensor has two parts: a capacitive humidity sensor and a thermistor. Additionally, there is an analog-to-digital conversion chip that converts the sensor readings to a digital signal.

The signal is fairly easy to read using any microcontroller, but reading temperature or humidity can take 250 milliseconds, and the reading you receive from the sensor may be up to two seconds behind the current temperature and humidity due to the slow read-out time.

However, the DHT-22 should operate just fine between −40 and 125°C with a ±0.5°C accuracy, and between 0 and 100% humidity with an accuracy of between 2 and 5%. It can be powered with 3.3V or 5V, and will draw 2.5mA at peak.

Wiring the Breadboard

Let's go ahead and start building. You'll need to grab your Arduino board, a breadboard, the DHT-22 sensor, a 10kΩ resistor, and some jumper wires.

It's standard practice to use red wires for power (VCC), which in our case is +5V, and black wires for ground (GND). This helps out a lot later on when you're struggling with a breadboard full of components and trying to figure out where all the wires are going.

Connect the +5V pin on your Arduino to the +ve rail of the breadboard, and one of the ground (GND) pins to the −ve rail. Next, plug the DHT-22 into the breadboard such that each pin is in a separate row (rather than on the same row, which would mean all the pins would be electrically connected to each other), and orientated it so that the front side of the sensor is facing towards the middle of the breadboard (see Figure 2-3).

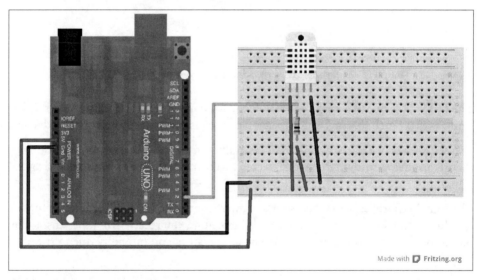

Figure 2-3. The wiring diagram

The DHT sensor has four pins. From left to right, these are: VCC (can be between +3 and +5V), data out, an unused pin (sometimes connected to GND) and GND.

We need to get power to our sensor. So connect one jumper wire from the +5V rail of the breadboard to the row that has the leftmost pin (PIN 1) of the DHT-22 sensor, and another from the GND rail of the breadboard to the rightmost pin (PIN 4) of the sensor.

There have been reports that some sensors have pin 3 and 4 reversed, possibly as the result of a manufacturing issue. If you have issues, try grounding both pins 3 and 4.

Then connect a jumper wire from PIN 2 of the Arduino board to the second pin from the left (PIN 2) on the DHT-22 sensor. This is the wire our data is going to be flowing over. We're also going to have to connect a *pull-up resistor* to this pin.

A pull-up resistor weakly "pulls" the voltage of the wire towards a known voltage level. The pull-up resistor brings the wire up to the high logic level.

In digital circuits, a logic level is one of a finite number of states a signal can have. They are usually represented by voltage levels in the circuit. In our case; +5V represents a **HIGH** logic level, and 0V (GND) represents a **LOW** logic level. By extension, these represent the digital 1 and 0, or on and off.

When our sensor is active, it will override the high logic level set by the pull-up resistor. The pull-up resistor ensures that the wire is at a defined logic level even if the sensor connected to it isn't active. A pull-down resistor works in the same way, but is connected to ground. It holds the logic signal near zero volts when no other active sensor is connected.

Insert a 10kΩ resistor so that it bridges the gap between the two sides of the breadboard, connecting the data pin (PIN 2) to the other side of the gap, then a jumper from that row to the +5V rail (see Figure 2-3 again).

Resistor Color Coding

A color code is normally used to denote the resistance value of a resistor. Normally, there will be four bands: three on one side to tell you what the resistance of the component is, and then a gap, and another band off to the side to tell you the tolerance of the component. In some cases, there may be an additional band as part of the resistor value, but it's fairly rare.

Orientate the resistor such that the band that is separated from the others is on the right, and then read the color bands left to right. In the case of our 10kΩ, we see that the bands are *BROWN, BLACK, ORANGE*. This translates to 1, 0 (so 10) and then a multiplier of 1kΩ (hence 10kΩ).

To the right of this is a *GOLD* band, which means that our resistor is 10kΩ ± 5%.

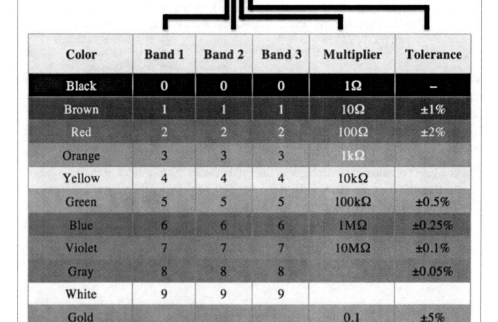

Color	Band 1	Band 2	Band 3	Multiplier	Tolerance
Black	0	0	0	1Ω	--
Brown	1	1	1	10Ω	±1%
Red	2	2	2	100Ω	±2%
Orange	3	3	3	1kΩ	
Yellow	4	4	4	10kΩ	
Green	5	5	5	100kΩ	±0.5%
Blue	6	6	6	1MΩ	±0.25%
Violet	7	7	7	10MΩ	±0.1%
Gray	8	8	8		±0.05%
White	9	9	9		
Gold				0.1	±5%
Silver				0.01	±10%

If you've followed the instructions and wired the breadboard as laid out in Figure 2-3, you should have something that looks a lot like Figure 2-4.

Figure 2-4. The Arduino and breadboarded DHT-22 sensor

Now we're ready to write the software that goes with the hardware.

Writing the Software

There are a number of Arduino libraries that provide extra functionality (and usually act as a convenience layer) for your sketches. Some libraries ship with the development environment, but you can also install your own third-party libraries.

Libraries make things simpler when dealing with commonly encountered hardware, and handle a lot of the heavy lifting behind the scenes—allowing you to focus on what you want to do with your hardware (and getting the data you want), rather than writing software to talk to it.

> Because libraries are uploaded to the board with your sketch, they increase the amount of space used by the sketch on the board. If a sketch no longer needs a library, simply delete its `#include` statements from the top of your code. This will stop the Arduino IDE from linking the library with your sketch and decrease the amount of space used on the Arduino board by your sketch.

The DHT Library

We're going to make use of the DHT library written by Limor Fried. It's available for download from the Adafruit GitHub repository (*https://github.com/adafruit/DHT-sensor-library*). The library supports not just the DHT-22 we're using here, but also the other sensors in the range: the DHT-11, DHT-21, and AM2301.

If you're familiar with Git, you can just check out the library from the Adafruit's GitHub page, but if not, go ahead and download the library as a ZIP file. You'll need to make sure the folder containing the library is named *DHT*, and you should rename it to that if necessary. It should contain the *DHT.cpp* and *DHT.h* files. Place the *DHT* library folder into your Arduino libraries folder (see the next sidebar for details), and remember to **quit the Arduino development environment** before installing the library.

Installing New Arduino Libraries

If you need to install a new library, as we do here, you should **quit the Arduino development environment** before going ahead and creating a libraries folder inside your default sketch location. You can check what this folder is called by looking at the Arduino→Preferences... menu.

By default, the new folder would be in *~/Documents/Arduino/libraries/* on OS X, while on Windows it would be *My Documents\Arduino\libraries*.

To add a library, create a new directory inside the libraries directory with the same name as your library. The folder should contain a C or C++ file with your code, and a header file with your function and variable declarations.

Once you've done this, **restart the Arduino development environment**. To use the library in a sketch, simply go to Sketch→Import Library, and choose it from the list of available libraries.

This will insert an `#include` statement at the top of your sketch for each header file in the library's folder. These statements make the public functions and constants defined by the library available to your sketch. They also signal the Arduino environment to link that library's code with your sketch when it is compiled or uploaded.

Arduino Sketch

Now that we've installed the DHT library, we can **reopen the Arduino development environment** and start writing some code.

You can import the DHT library either by using the Sketch→Import Library dropdown menu, or by adding the `#include` line by hand to your sketch. Then go ahead and enter the following code:

```
#include <DHT.h> ❶
#define DHTTYPE DHT22 ❷

int dhtPin = 2; ❸
DHT dht(dhtPin, DHTTYPE);

void setup() {
    pinMode(dhtPin, INPUT);     // declare DHT sensor pin as input
    Serial.begin(9600);         // open the serial connection to your PC
    dht.begin();
}

void loop() {
  float h = dht.readHumidity(); ❹
  float t = dht.readTemperature();

  if (isnan(t) || isnan(h)) {
    Serial.println("Error: Failed to read from DHT");
  } else {
    Serial.print( "T = " ),
    Serial.print( t );
    Serial.print( "C, H = " );
    Serial.print( h );
    Serial.println( "%" );
  }
  delay(2000);

}
```

❶ We need to remember to import the Adafruit DHT library

❷ The DHT library actually supports the DHT-22, DHT-11, DHT-21, and AM2301 sensors. Here, we're telling it which of the four supported sensors we're using.

❸ Defining the pin to which the DHT data pin is connected; see Figure 2-3.

❹ The DHT library provides two routines that readTemperature( ) and readHumidity( ). By default, readTemperature( ) will return the temperature in centigrade, however by passing a 1 we can make it return Fahrenheit.

Once you've entered the code, plug your Arduino into your computer, then go ahead and compile and upload the sketch to your board.

See the sections "Connecting to the Board" on page 5 and "Uploading the Sketch" on page 10 in Chapter 1 if you're having trouble uploading the sketch to your Arduino board.

When you see the "*Done uploading.*" message in the status bar, click on the Serial Console button in the development environment to open the serial console.

Running the Software

Opening the serial console will reset the Arduino, at which point you should see something a lot like Figure 2-5. Every couple of seconds, a new reading of the temperature and humidity will be printed to the console.

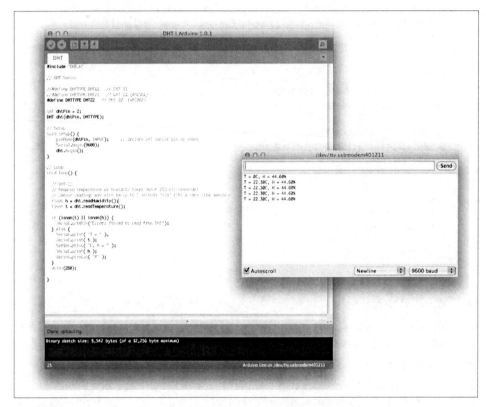

Figure 2-5. The Arduino should report the temperature and humidity in the Serial Console

Right now, getting to your data is pretty hard, and later in the book we're going to look at making that easier as well as creating some basic visualizations. For now, let's build our hardware platform first. However, if you keep the code running long enough, you should see that the temperature and humidity change over time (see Figure 2-6).

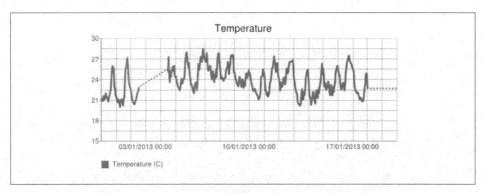

Figure 2-6. Temperature against time (dotted lines denote gaps in the data)

Summary

In this chapter we've learned how to wire up a breadboard to prototype a circuit, and how use the Arduino to read values from a sensor.

CHAPTER 3
Adding Another Sensor

Just as when writing code, taking an iterative approach to building hardware is always advisable: change one thing, test, and then change another. In later chapters, we'll take the evolutionary approach one more step forward and move our sensors off the breadboard, but for now, let's add another sensor.

The Sensor

The next sensor we'll add is one to detect motion.

PIR Sensor

We're going to make use of a Passive Infra-Red (PIR) motion sensor (see Figure 3-1).

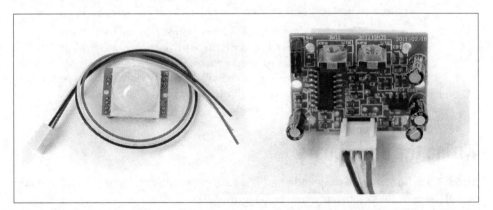

Figure 3-1. A typical PIR sensor, front (left) and rear (right) with connector and potentiometers to adjust the sample rate and sensitivity of the sensor

PIR sensors are commonly used in alarm systems, as they're an inexpensive and fairly easy-to-use way to detect motion. While there are a number of different PIR boards commonly available (and each will have slightly different specifications), they all pretty much work the same way.

A PIR consists of a pyroelectric sensor that can detect levels of Infra-Red (IR) radiation split into two halves. The halves are wired to cancel each other out if the amount of IR radiation is constant. However, if the levels are changing, such that one half of the sensor sees more (or less) than the other, then the combined output of the two will be pulled high or low.

The PIR board in Figure 3-1 contains a pyroelectric sensor underneath a Fresnel lens (the plastic dome on the front of the board), along with supporting circuitry and a chip dedicated to taking the analog output of the sensor and emitting a digital output pulse.

Connecting most PIR sensors to a microcontroller is really simple. The PIR acts as a digital output, so all you need to do is listen for the pin on the Arduino: when the pin is low, there is no motion, and when the pin is pulled high by the PIR sensor, it has detected motion.

 Most PIR sensors will have two potentiometers on the rear; in Figure 3-1, these are the orange knobs towards the top of the board shown on the right. These are used to adjust the sensitivity to movement, and the refresh time of the sensor. When triggered, the PIR will stay "high" for a certain time period determined by adjusting the potentiometer. Movement during this time won't retrigger the PIR. However, when the PIR drops "low," it will be retriggered again if there is still ongoing movement at that time. Effectively, this second potentiometer controls the time-domain resolution of the PIR sensor.

Also on the back is a small black jumper (top left on the back of the board), which determines whether the PIR is in "re-trigger" mode. Most PIR boards will ship with this jumper enabled. If your PIR sensor is not in re-trigger mode, the sensor output will drop low every second or so as the onboard controller chip polls the sensor. For most applications, you'll want to be in re-trigger mode.

Wiring the Breadboard

Building on both our existing code and our existing hardware, lets go ahead and add our PIR sensor to our current setup (see Figure 3-2).

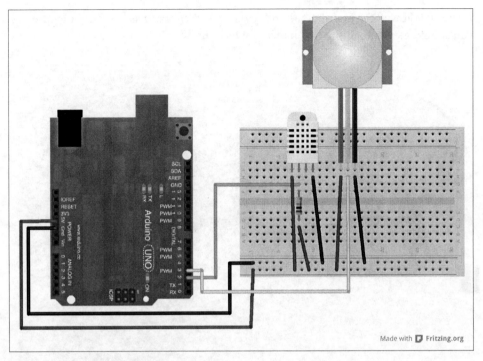

Figure 3-2. The wiring diagram

Look at the silkscreen on the back of the PIR sensor. The three pins on the male socket should be labeled: GND, OUT, and +5V. Plug the connector on the *trailing leads* (the wires that come with the sensor) into the socket on the back of the sensor board, making sure the black wire is plugged into the GND pin. Then go ahead and plug the trailing leads into the breadboard (see Figure 3-2).

 You might have to use a pair of adjustable wire strippers—or, failing that, some flush diagonal cutters—to strip the ends of the trailing leads before inserting them into the breadboard. If the leads on your PIR sensor are stranded wire, you may have to *tin* the ends of the wires (coat them with a small amount of solder) with a soldering iron before inserting them into the breadboard. See Chapter 5 for more information about how to solder if you haven't done any soldering before.

Connect the red trailing lead to the +ve (+5V) rail using a jumper wire, and the black trailing lead to the −ve (GND) rail using another jumper wire. Finally, connect the middle (yellow) wire that carries the data signal to PIN 3 of your Arduino with another jumper wire.

If you've followed the instructions and wired the breadboard as laid out in Figure 3-2, you should have something that looks a lot like Figure 3-3.

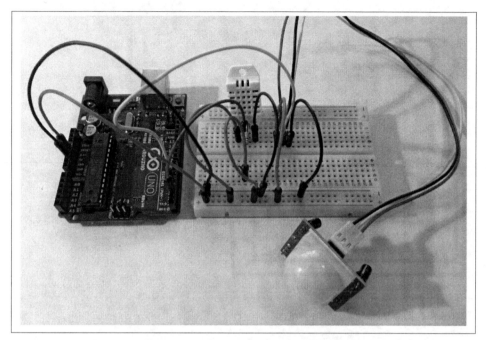

Figure 3-3. The Arduino and breadboarded DHT-22 and PIR sensors

Now that we've modified our hardware, we're ready to modify the software that goes with it.

Modifying the Software

Reading from the PIR motion sensor is a lot easier than reading from our DHT temperature and humidity sensor. We won't need an external library. Instead, we just need to examine the state of the digital pin to which we've connected the sensor on the Arduino board:

```
#include <DHT.h>
#define DHTTYPE DHT22

int dhtPin = 2;
DHT dht(dhtPin, DHTTYPE);

int pirPin = 3;
int pirState = LOW;    // we start, assuming no motion detected
int pirVal = 0;
int motionState = 0;
```

```
void setup() {
    pinMode(dhtPin, INPUT);    // declare DHT sensor pin as input
    pinMode(pirPin, INPUT);    // declare PIR sensor pin as input

    Serial.begin(9600);        // open the serial connection to your PC
    dht.begin();
}

void loop() {
  float h = dht.readHumidity();
  float t = dht.readTemperature();

  if (isnan(t) || isnan(h)) {
    Serial.println("Error: Failed to read from DHT");
  } else {
    Serial.print( "T = " ),
    Serial.print( t );
    Serial.print( "C, H = " );
    Serial.print( h );
    Serial.println( "%" );
  }

  pirVal = digitalRead(pirPin);  // read input value
  if(pirVal == HIGH){            // check if the input is HIGH
    if(pirState == LOW){
      // we have just turned on
      motionState = 1;
      Serial.println( "Motion started" );❶

      // We only want to print on the output change, not state
      pirState = HIGH;
    }
  } else {
    if(pirState == HIGH){
      // we have just turned of
      motionState = -1;
      Serial.println( "Motion ended" );❷

      // We only want to print on the output change, not state
      pirState = LOW;
    }
  }

  motionState = 0; // reset motion state
  delay(2000);

}
```

❶ The pin to which the PIR sensor is connected has been pulled **HIGH** by the sensor board, and last time through our loop the state was **LOW**. This means that since the last time we looked, movement has started.

Modifying the Software | 31

❷   The pin the PIR sensor is connected to is **LOW**, however last time we looked it was **HIGH**. This means that whatever movement the PIR detected has now ended.

Once you've entered the code, plug your Arduino into your computer, then go ahead and compile and upload the sketch to your board.

See the sections "Connecting to the Board" on page 5 and "Uploading the Sketch" on page 10 in Chapter 1 if you're having trouble uploading the sketch to your Arduino board.

When you see the "*Done uploading.*" message in the status bar, click on the Serial Console button in the development environment to open the serial console.

Running the Software

Opening the serial console will reset the Arduino, at which point you should see something a lot like Figure 3-4. Every couple of seconds, a new reading of the temperature and humidity will be printed to the console. Interspaced with this will be notifications from the PIR sensor of the beginning and end of any movement.

The PIR sensor will probably take five to ten seconds to "warm up" before it can detect movement. Waving your hand in front of it before that won't trip the sensor.

Figure 3-4. The Arduino should report the temperature and humidity in the Serial Console along with "motion started" and "motion ended" notifications

Summary

In this chapter we've added another sensor to our hardware, and extended our code to deal with it. Just as you build basic functionality and then extend it when writing software, it's important to follow the same principal when building hardware.

CHAPTER 4
Finishing the Breadboard

Now that our sensor platform both measures the current environmental conditions and looks out for motion, let's add our final sensor and finish building our prototype.

The Sensor

The next sensor we'll add is one to detect changes in ambient noise levels.

Electret Microphone

We're going to use the electret microphone breakout board from Adafruit Industries (see Figure 4-1). The board comes with a capsule microphone and supporting hardware including, an op-amp and a potentiometer to allow you to adjust the gain.

Electret microphones are small, omnidirectional microphones commonly used in cell phones and laptops. They can have an extremely wide frequency response; the one we're using has a 20Hz to 20kHz frequency response. They are small, very cheap to produce, and quite sensitive. Unfortunately, they also have some drawbacks, including a very uneven frequency response, high noise floor, and high levels of distortion. Despite that, they're a good choice if fidelity isn't an overriding issue for your application.

Since the electret microphone only produces a few milli-volts peak-to-peak, it needs to be coupled with an op-amp before it can be used. The Adafruit breakout board uses a Maxim MAX4466 for this purpose.

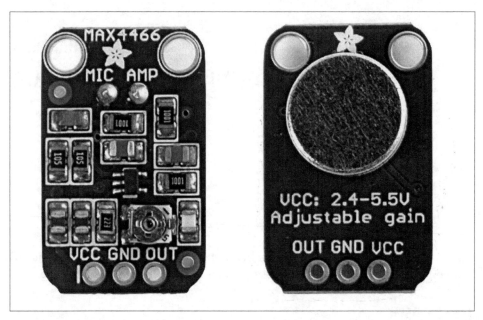

Figure 4-1. The Adafruit Electret Microphone Amplifier back with potentiometer (left) and front with the transducer (right)

Wiring the Breadboard

Starting where we left off at the end of the last chapter, let's go ahead and add the microphone to our current setup (see Figure 4-2).

Looking at the silkscreen on the Adafruit board, you'll see that there are three connectors (see Figure 4-2 again). Looking at the front, the output (signal) is on the left, labeled OUT; the middle connector is the ground, labeled GND; and the power is on the right, labeled VCC. You'll also see from the silkscreen that we can drive the board with anything from 2.4V up to 5.5V. That's good, as our Arduino microcontroller board runs at 5V. We can plug the microphone board directly into the Arduino.

Connect a jumper wire from the +ve (+5V) rail to the VCC connector of the breakout board, and another from the –ve (GND) rail to the GND connector. Finally, connect another jumper wire from the OUT connector on the microphone breakout board to the A0 pin on the Arduino.

 The A0 pin on the Arduino is on the lefthand side of the board when you're looking at it with the USB and power jacks facing away from you. While you can use this pin as a normal digital input/output pin, this is a special pin that can also be used for analog input.

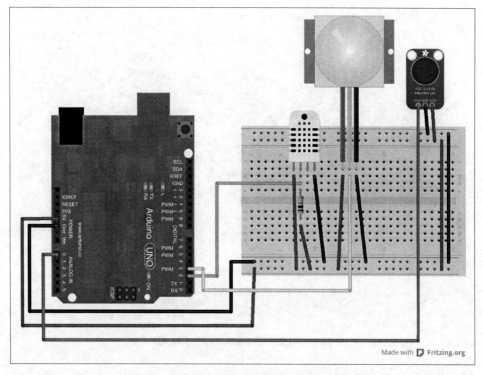

Figure 4-2. The wiring diagram for incorporating the electret microphone

You may notice in Figure 4-3 that we've actually wired the breadboard slightly differently from the wiring diagram in Figure 4-2. This is because our microphone breakout board came with three header pins, and instead of connecting the wires to the microphone board, we soldered the headers to it and then plugged that into the breadboard. This really simplified things.

 See Chapter 5 for more information about how to solder if you haven't done any soldering before. If you don't want to solder the header pins to the microphone board quite yet, you can probably get away with threading some jumper wires through the holes and firmly wrapping them around the board. You'll need to make sure that they make a good connection to the pad, and that they don't touch and short each other out.

If you've followed the instructions, you should have something that looks a lot like Figure 4-3.

Figure 4-3. The Arduino and breadboarded DHT-22, PIR, and microphone

Modifying the Software

Calibrating sound level readings so that you get a value measured in decibels is actually a really hard to do accurately, so we're not even going to try here. In any case, we're really only interested in knowing how the noise level around the sensor platform changes over time, rather than the absolute value of the sound pressure on the microphone.

We can determine this far more easily than measuring an absolute (calibrated) value by keeping a running average of the sound pressure on the microphone and reporting how the current reading has changed with respect to our running average.

Go ahead and make the changes highlighted in bold below to our code:

```
#include <DHT.h>
#define DHTTYPE DHT22
#define SILENT_VALUE 380 // starting neutral mic value (self-correcting) ❶

int dhtPin = 2;
DHT dht(dhtPin, DHTTYPE);

int pirPin = 3;
int pirState = LOW;    // we start, assuming no motion detected
int pirVal = 0;
int motionState = 0;
```

```
int micPin = 0;
int micVal = 0;

void setup() {❷
    pinMode(dhtPin, INPUT);    // declare DHT sensor pin as input
    pinMode(pirPin, INPUT);    // declare PIR sensor pin as input

    Serial.begin(9600);        // open the serial connection to your PC
    dht.begin();
}

void loop() {
  float h = dht.readHumidity();
  float t = dht.readTemperature();

  if (isnan(t) || isnan(h)) {
    Serial.println("Error: Failed to read from DHT");
  } else {
    Serial.print( "T = " ),
    Serial.print( t );
    Serial.print( "C, H = " );
    Serial.print( h );
    Serial.println( "%" );
  }

  pirVal = digitalRead(pirPin);  // read input value
  if(pirVal == HIGH){            // check if the input is HIGH
    if(pirState == LOW){
      // we have just turned on
      motionState = 1;
      Serial.println( "Motion started" );

      // We only want to print on the output change, not state
      pirState = HIGH;
    }
  }else{
    if(pirState == HIGH){
      // we have just turned of
      motionState = -1;
      Serial.println( "Motion ended" );

      // We only want to print on the output change, not state
      pirState = LOW;
    }
  }

  micVal = getSound();
  Serial.print( "MIC = " );
  Serial.println( micVal );

  motionState = 0; // reset motion state
```

```
    delay(2000);

}

int getSound() {
  static int average = SILENT_VALUE;❸
  static int avgEnvelope = 0;❹
  int avgSmoothing = 10;❺
  int envSmoothing = 2;
  int numSamples = 1000;
  int envelope = 0;❻
  for ( int i = 0; i< numSamples; i++ ) {
    int sound = analogRead(micPin);❼
    int sampleEnvelope = abs(sound - average);
    envelope = (sampleEnvelope+envelope)/2;
    avgEnvelope = (envSmoothing * avgEnvelope + sampleEnvelope) /
                  (envSmoothing + 1);
    average = (avgSmoothing * average + sound) / (avgSmoothing + 1);
  }
  return envelope;
}
```

❶ The analogRead() command converts the input voltage range, 0 to 5 volts, to a digital value between 0 and 1023. Here we're setting the "silent" value to be 380, or around 1.85V. This "background noise" level will self correct over time.

❷ Since the microphone is producing an analog signal, we don't have to initialize the A0 pin for either input or output. This isn't necessary when we're using the pin in analog mode, because in this mode they can be only be used for input. Without additional hardware, the Arduino cannot produce an analog signal, although it can fake it using Pulse Width Modulation (PWM) on some pins. These pins are marked with a ~ symbol on the silkscreen.

❸ The running average is where the current neutral position for the microphone is stored.

❹ The average envelope level is the running average for the sound pressure level.

❺ Larger values of aveSmoothing give more smoothing for the average, while larger values for envSmoothing give more smoothing for the envelope.

❻ The envelope is the mean sound taken over many samples.

❼ Instead of digitalRead( ) we instead use analogRead( ) to get the state of the A0 pin to which our microphone connected. As mentioned above, this will return a digital value between 0 and 1023 representative of the analog voltage level present on the pin.

Once you've entered the code, plug your Arduino into your computer, then go ahead and compile and upload the sketch to your board.

See the sections "Connecting to the Board" on page 5 and "Uploading the Sketch" on page 10 in Chapter 1 if you're having trouble uploading the sketch to your Arduino board.

When you see the "*Done uploading.*" message in the status bar, click on the Serial Console button in the development environment to open the serial console.

Running the Software

Opening the Serial Console will reset the Arduino, at which point you should see something a lot like Figure 4-4. Every couple of seconds, a new reading of the temperature, humidity, and (sound) envelope values will be printed to the console. As before, interspaced with these will be the notifications from the PIR sensor of the beginning and end of any movement.

Figure 4-4. The Arduino should now also report noise levels in the serial console

Remember that the envelope value being reported is not a measurement of the absolute volume of the noise (sound pressure) on the microphone, but instead is a measurement of the change in this noise.

Adding Some LEDs

While we have haven't made use of them much so far, as our Arduino boards have been connected directly to our laptop and we've been able to use the serial connection to see what's going on, LEDs are used almost ubiquitously as a debugging tool when building hardware. Later in the book (see Chapter 7), we're going to unhook our Arduino from our laptop and put it on the network, and we won't be able to see the serial output from the board directly any more. So now is a good time to add some LEDs to our project to help us with debugging later on.

We're going to use three LEDs (see Figure 4-5): one to show that our board has booted and is turned on, another to show us that it's running our loop( ) correctly, and a final one to show us if there has been any movement detected by the PIR sensor.

 I'm going to use a green LED for power, and two red LEDs for the loop and motion indicator lights. You don't have to do that; any LED will do.

Insert three LEDs into your breadboard as shown in Figure 4-5, and use a jumper wire to connect the GND pin of each of the LEDs to the −ve (GND) rail of the breadboard.

 Remember that LEDs are directional components and must not be inserted backwards. Look carefully at the two legs of the LED; one should be shorter than the other. The shorter leg corresponds to the ground, while the longer leg is the positive.

Then, in a similar manner to the way we wired the pull-up resistor we used for the DHT-22 sensor, connect a 220Ω resistor to the positive pin of each of the three LEDs, bridging the gap between the two sides of the breadboard. Then use a jumper wire to connect the three LEDs to PINs 13, 9, and 8 on the Arduino board, respectively going left-to-right (see Figure 4-5 again if needed).

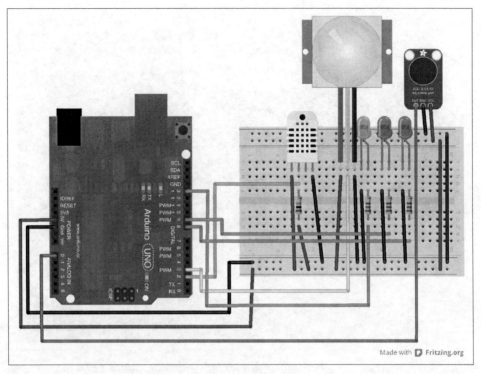

Figure 4-5. The wiring diagram with added LEDs

 A resistor is needed inline between the LED and the Arduino to limit the current flowing across the resistor; if you drive an LED with too high a current, you will destroy it. If you think back to Chapter 1, you'll remember that we plugged an LED directly into PIN 13 of the Arduino board. While it's generally okay to do that, it's not good practice, and you should avoid it now that you know better.

If you've followed the instructions and the wired the breadboard as laid out in Figure 4-5, you should have something that looks a lot like Figure 4-6.

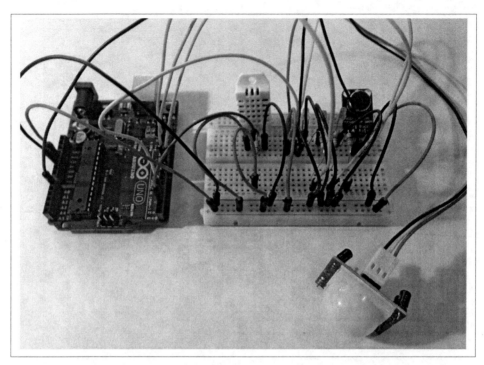

Figure 4-6. The Arduino and breadboarded sensors with some added LEDs

Now we're changed the hardware, it's time to modify our software.

Modifying the Software

The changes to our Arduino sketch are pretty self-explanatory. We'll set up the three LED pins as `OUTPUT` in the `setup()` function and then toggle them **HIGH** and **LOW** depending on what we want to indicate.

Make the following changes shown in bold to your code:

```
#include <DHT.h>
#define DHTTYPE DHT22
#define SILENT_VALUE 380   // starting neutral microphone value (self-correcting)

int dhtPin = 2;
DHT dht(dhtPin, DHTTYPE);

int pirPin = 3;
int pirState = LOW;    // we start, assuming no motion detected
int pirVal = 0;
int motionState = 0;
```

```
int micPin = 0;
int micVal = 0;

int powPin = 13;
int ledPin = 8;
int motPin = 9;

void setup() {
    pinMode(dhtPin, INPUT);      // declare DHT sensor pin as input
    pinMode(pirPin, INPUT);      // declare PIR sensor pin as input

    pinMode(powPin, OUTPUT);     // Power LED
    pinMode(ledPin, OUTPUT);     // Rx/Tx LED
    pinMode(motPin, OUTPUT);     // Motion Detected LED

    Serial.begin(9600);          // open the serial connection to your PC
    dht.begin();
    digitalWrite(powPin, HIGH);

}

void loop() {
  digitalWrite(ledPin, HIGH);

  float h = dht.readHumidity();
  float t = dht.readTemperature();

  if (isnan(t) || isnan(h)) {
    Serial.println("Error: Failed to read from DHT");
  } else {
    Serial.print( "T = " ),
    Serial.print( t );
    Serial.print( "C, H = " );
    Serial.print( h );
    Serial.println( "%" );
  }

  pirVal = digitalRead(pirPin);  // read input value
  if(pirVal == HIGH){            // check if the input is HIGH
    if(pirState == LOW){
      // we have just turned on
      motionState = 1;
      digitalWrite(motPin, HIGH);  // turn LED ON
      Serial.println( "Motion started" );


      // We only want to print on the output change, not state
      pirState = HIGH;
    }
  }else{
    if(pirState == HIGH){
      // we have just turned off
```

Adding Some LEDs | 45

```
      motionState = -1;
      digitalWrite(motPin, LOW);   // turn LED OFF
      Serial.println( "Motion ended" );

      // We only want to print on the output change, not state
      pirState = LOW;
    }
  }

  micVal = getSound();
  Serial.print( "MIC = " );
  Serial.println( micVal );

  motionState = 0; // reset motion state
  digitalWrite(ledPin, LOW);
  delay(2000);

}

int getSound() {
  static int average = SILENT_VALUE;
  static int avgEnvelope = 0;
  int avgSmoothing = 10;
  int envSmoothing = 2;
  int numSamples=1000;
  int envelope=0;
  for (int i=0; i<numSamples; i++) {
    int sound=analogRead(micPin);
    int sampleEnvelope = abs(sound - average);
    envelope = (sampleEnvelope+envelope)/2;
    avgEnvelope = (envSmoothing * avgEnvelope + sampleEnvelope) /
                  (envSmoothing + 1);
    average = (avgSmoothing * average + sound) / (avgSmoothing + 1);
  }
  return envelope;
}
```

Once you've entered the code, plug your Arduino into your computer, then go ahead and compile and upload the sketch to your board.

See the sections "Connecting to the Board" on page 5 and "Uploading the Sketch" on page 10 in Chapter 1 if you're having trouble uploading the sketch to your Arduino board.

When you see the "*Done uploading.*" message in the status bar, click on the Serial Console button in the development environment to open the serial console.

Running the Software

If you open up the serial console in the Arduino development environment, everything should pretty much be exactly the same as before. However, this time there should be some blinking lights to accompany your data (see Figure 4-7).

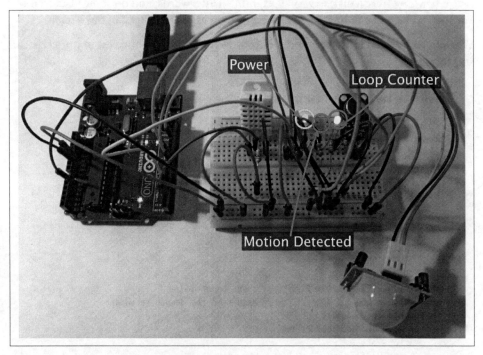

Figure 4-7. The LEDs show the current state of the sketch, and are helpful for monitoring our board when we unplug it from our laptop

The leftmost LED should turn on at the end of the setup() function and stay on: this is the LED we're using to indicate that the power to the board is on and our sketch has started. The rightmost LED should be pulled high (and thus turn on) at the start of the loop() function, and then be pulled low (and thus turn off) after the loop completes. This means that the LED will slowly blink on and off, showing us that data is being taken. Finally, the middle LED should be pulled high (and turn on) when the PIR sensor detects the start of motion, and then low again (and turn off) at the end of motion.

Making the Output Machine-Readable

Right now the output of our sketch to the serial port was designed to be read by a human. However, in the next chapter we're going use some Python code to collect our data and save it to a CSV file on disk so we can examine it afterwards, and do some visualization. For that, we're going to have to make some changes to the output of our sketch.

We only really need to modify serial output in the loop( )—the rest of our code can stay the same:

```
void loop() {
  digitalWrite(ledPin, HIGH);

  float h = dht.readHumidity();
  float t = dht.readTemperature();

  if (isnan(t) || isnan(h)) {
    // Don't output temperature and humidity readings
  } else {
    Serial.print( t );
    Serial.print( ", " );
    Serial.print( h );
    Serial.print( ", " );
  }

  pirVal = digitalRead(pirPin);   // read input value
  if(pirVal == HIGH){             // check if the input is HIGH
    if(pirState == LOW){
      // we have just turned on
      motionState = 1;
      digitalWrite(motPin, HIGH);   // turn LED ON

      // We only want to print on the output change, not state
      pirState = HIGH;
    }
  }else{
    if(pirState == HIGH){
      // we have just turned off
      motionState = -1;
      digitalWrite(motPin, LOW);   // turn LED OFF

      // We only want to print on the output change, not state
      pirState = LOW;
    }
  }

  micVal = getSound();
  Serial.print( micVal );
  Serial.print( ", " );
```

```
    Serial.print( pirVal );
    Serial.print( ", " );
    Serial.println( motionState );

    motionState = 0; // reset motion state
    digitalWrite(ledPin, LOW);
    delay(2000);

}
```

After making these changes, re-upload the sketch to your hardware. If you reopen the serial console, the output from the sketch should be changed from this:

```
T = 22.60C, H = 42.50%
MIC = 14
T = 22.50C, H = 42.60%
MIC = 25
Motion Started
T = 22.50C, H = 42.60%
MIC = 25
T = 22.60C, H = 42.60%
MIC = 30
T = 22.70C, H = 42.70%
MIC = 114
Motion Ended
T = 22.60C, H = 42.70%
MIC = 20
```

to something more like this:

```
22.60, 42.50, 14, 0, 0
22.50, 42.60, 25, 1, 1❶
22.60, 42.60, 30, 1, 0
22.70, 42.70, 114, 1, -1
22.60, 42.70, 20, 0, 0
```

❶ Here we're using the previously unused `motionState` variable to show the start and end of a PIR motion event. A 1 denotes the start of motion, and a -1 denotes the end of motion.

…as we can see in Figure 4-8.

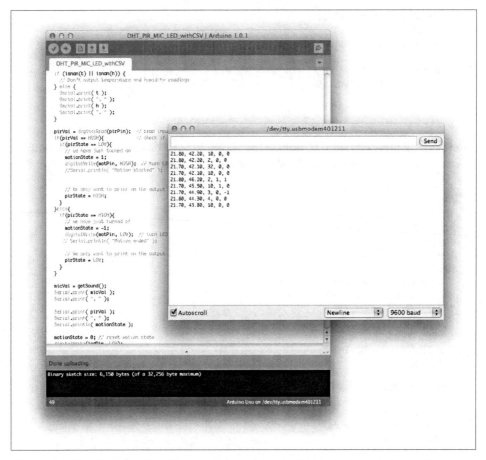

Figure 4-8. The Arduino should now also report noise levels in the serial console

This will be a lot easier to read into Python—or any other sort of code—than the more human-friendly but hard-to-parse output we had before.

Communicating with Python

We will use Python as a quick tool to talk to our Arduino. If you haven't used Python before, there are many tutorials and books that will help get you started. Our intention here is to give you the code required to connect to the Arduino and start receiving data.

First, make sure you have the pySerial module installed on your computer. Although the Arduino is connected via a USB port, it uses the computer's serial interface protocols for communication.

To talk to the Arduino, we need to find the port that the Arduino is connected to. On a typical Macintosh, the Arduino will show up as a USB modem device:

```
%python -m serial.tools.list_ports
/dev/tty.Bluetooth-Modem
/dev/tty.Bluetooth-PDA-Sync
/dev/tty.usbmodem621
3 ports found
%
```

Your ports may not look exactly like this, but should be similar. Once you have found a USB modem device that looks promising, use the `ser` command to open the serial port in order to talk to the Arduino. Make sure you have first imported the serial library into Python.

```
>>> import serial
>>> ser = serial.Serial('/dev/tty.usbmodem621')
```

If you get any errors, try another one of the serial ports. If that doesn't work, make sure no other devices are currently using the serial port in question. Unplug and then reconnect your Arduino to make sure the port is not in use. Once you have successfully opened the serial port, the following quick test will confirm a working connection:

```
>>> x = ser.read(30)
>>> print x
```

If you are running the demo code from the sensor mote project, you should see a printout that looks like this:

```
22.60, 42.50, 14, 0, 0
```

After you have tested your connection using the Python command line, feel free to close the serial port:

```
>>> ser.close()
```

We can take our Python code a step further. The following Python script will collect time-stamped data from your sensor mote and save it to a *.csv* file:

```
import serial
import time

ser = serial.Serial('/dev/tty.usbmodem621', 9600, timeout=1)
time.sleep(1)

logfile = open('test.csv', 'a')

count = 0
while (count < 5):
    count = count + 1
    line = ser.readline()
    now = time.strftime("%d/%m/%Y %H:%M:%S", time.localtime())
    a =  "%s, %s, %s" % (now, line, "\n")
    print a
    logfile.write(a)
    logfile.flush()
```

```
logfile.close()
ser.close()
```

Once you have confirmed that you can collect data and send it to a log file, you have completed the steps required to build a sensor mote by hand and connect it to your computer.

 If you are interested in being able to move your data from the Arduino to a TCP connection, this can be done simply by downloading and implementing a Python TCP to Serial bridge example (*http://pyserial.sourceforge.net/examples.html#single-port-tcp-ip-serial-bridge-rfc-2217*).

Summary

In this chapter we've added our final sensor to the breadboard, finalized our software and hardware design, and ran some simple tests with Python. In the next chapter we're going to move our project off the breadboard.

CHAPTER 5
Moving from Breadboard to Prototype

Building a circuit on a solderless breadboard is a great starting point for most projects. We've seen that solderless breadboards provide us with an easy way to mount devices and move pins around quickly. However, the ease with which wires can be added and removed is a problem when we want to put our sensor system out into the environment.

The Prototype

Once we have tested our circuit and gotten it working on a solderless breadboard, the next step is to build a prototype on which the wires and components are more permanently connected together.

We will use solder, the electronics equivalent of glue, to build our prototype. It is a metal alloy that melts at temperature low enough that it can be melted around the pins of our circuit components without causing damage to the sensitive electronics inside these components. When the solder cools, it creates a solid, electrically conductive, permanent bond to our wires and pins.

Before we make these permanent connections, however, it is very important to document the wiring of the circuit we designed on our solderless breadboard. Clear and easy-to-read documentation of our solderless breadboard is necessary to help us see what pins and wires to solder together as we build our prototype, especially if our prototype is complex. Good documentation will also allow us to share our circuit with others. This can be valuable during the debugging process. Finally, this documentation can be sent to a factory if we want to turn our prototype into a manufactured product. The easiest way to document our solderless breadboard is with a program called Fritzing.

Fritzing

Fritzing is a program that lets us convert our circuit design into a direct graphical representation on the computer. Its main benefit is that a circuit designed in Fritzing looks like the real thing. This will give us an intuitive, direct, and immediately recognizable representation of our solderless breadboard and the components we've connected to it. Fritzing is much better than the simple alternative of taking a photograph of the breadboard (Figure 5-1).

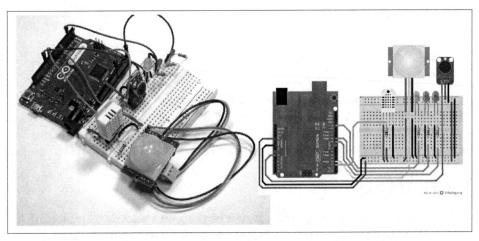

Figure 5-1. The real circuit is on the left; the Fritzing version is on the right

Fritzing lets us see wires, pins, and connections that may be obscured in a photo. There are many programs that we could use to document our circuit board. These programs, called *CAD* (Computer Aided Design) or *EDA* (Electronic Design Automation), all force us to document our circuit design using symbolic representations of the circuit components and connections. There is degree of training and familiarity required to know what symbol represents a particular component on our breadboard. Fritzing is the only program currently available which uses a direct graphical representation of the parts that we are using, so we can easily look at our Fritzing diagram and know how things are connected on our solderless breadboard; it was designed with the beginner in mind. Fritzing is free and available on the Mac, PC, and Linux.

As our circuits get more complex, we may need to use more sophisticated and expensive software for our circuit board design. But for now, we will focus on Fritzing as our starting point, and look at several features that make it a great tool for documenting your first breadboard.

Installing the Software

Download the latest version of the software from the Fritzing.org website.

 The latest version of the Fritzing development environment is available here (*http://fritzing.org/download/*). At the time of writing, this was Fritzing 0.7.11b.

Like most Mac applications, the development environment comes as a disk image (*.dmg* file) that should mount automatically after you have downloaded it. If it doesn't, double-click on it to open it manually. After it is open, just click-and-drag the application into your */Applications* folder.

Once you've installed the development environment, eject the disk image by dragging it to the trashcan and double-click on the *Fritzing.app* application icon to start the IDE. You should see something that looks a lot like Figure 5-2.

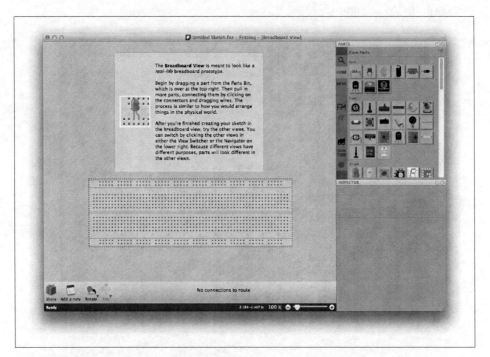

Figure 5-2. The Fritzing Breadboard View

Building a Fritzing Circuit

We'll begin our Fritzing circuit design by drawing the circuit in Fritzing exactly the way it looks in real life. When you first open Fritzing, you'll be presented with the Breadboard View. This view gives you the now-familiar breadboard front and center. Off in the upper-right corner of the Fritzing window is a parts bin, from which you can grab virtual circuit components and drag them into our circuit. A standard set of core parts are initially presented to us in the parts bin.

 Fritzing contains several parts bins from well-known suppliers like Adafruit, Arduino, Parallax, PicAxe, Sparkfun, and SnootLabs, in addition to the core parts.

We will begin our circuit design by adding the Arduino board to our project. Since the Core Parts bin does not include the Arduino boards, select the Arduino parts bin to add an Arduino to the circuit. Do this by clicking on the Arduino icon at the left edge of the parts bin, circled in Figure 5-3.

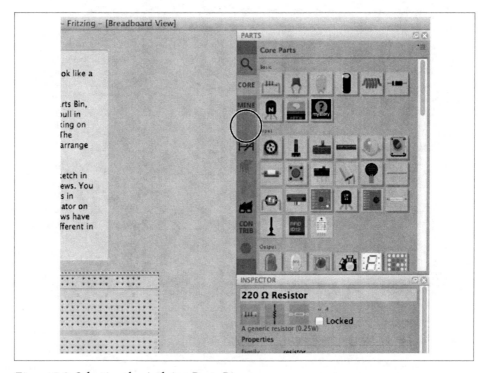

Figure 5-3. Selecting the Arduino Parts Bin

Once the Arduino parts bin is selected, you should see a dozen or more Arduino boards. Select the Arduino Uno R3 (or whichever other board you are using), and drop it anywhere into the breadboard view. Once you have added an Arduino to your circuit, you should have something that looks like Figure 5-4.

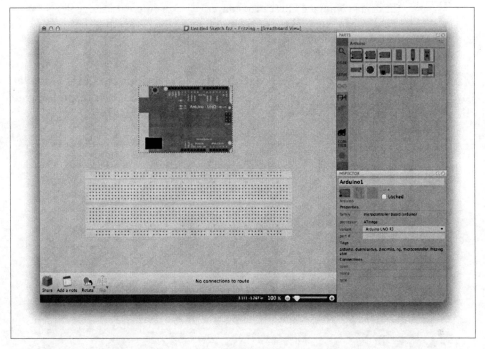

Figure 5-4. Our Fritzing breadboard with an Arduino Uno R3

 You may notice the parts inspector in the lower-right corner of the Fritzing window. Various attributes of the components can be viewed or selected in the Properties boxes in the inspector window. You can also select or deselect the "locked" checkbox. This locks a part down so that you don't accidentally move or delete it as you build your circuit.

Go ahead and select the "locked" option for both the Arduino and the breadboard. This will make it easier to add, move, and wire new parts.

Adding parts from other parts bins is as simple as clicking on that parts bin, and dragging the desired part onto the circuit. We can begin to add other components to the circuit, like LEDs and sensors.

Let's add the LEDs next. We want to add two 5mm red LEDs and one 5mm green LED. Return to the Core Parts bin in Fritzing, and scroll down until "Output" is visible. Select the red LED and drop three of them onto the breadboard.

 The default 5mm red LED can be converted into a different size or color LED by changing package or color options in the "Properties" box of the Parts Inspector.

Make sure the "locked" box is unchecked so that we can reposition the LEDs. Once you have done that, and changed the package and color options in the Parts Inspector, your circuit should look similar to Figure 5-5.

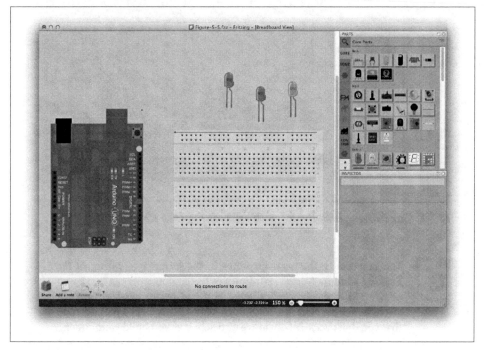

Figure 5-5. Fritzing diagram with LEDs

Adding Wires to the Diagram

Now we will plug the LEDs into the breadboard, and add wires to connect the LEDs to the Arduino.

First, select any of the LEDs. When not connected to anything, the tips of the legs will appear as red dots. When you drag your LED over the breadboard, the tips of the legs turn blue and a blue border appears around the LED when you are hovering over holes on the breadboard that the LED can plug into. If you release the LED over suitable holes, it will snap into place, and the tips of the legs will turn green, indicating that a connection has been made (see Figure 5-6).

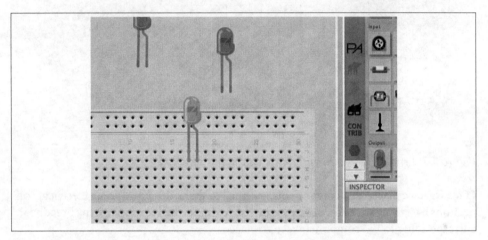

Figure 5-6. The LED is now connected to the breadboard

 You may notice that several of the vertically adjacent holes have turned green as well. This indicates that all of these circuit points are wired together. Anything we plug into these vertically adjacent holes will be connected to our LED.

Go ahead and drag all of the LEDs onto the breadboard. Repeat this process for the other components in the circuit as well. You will find the LEDs and resistors in the Core Parts bin. The PIR sensor and humidity sensor are located in the Adafruit parts bin.

If you've followed the instructions and placed all your parts onto the Fritzing circuit, it should look like the diagram in Figure 5-7.

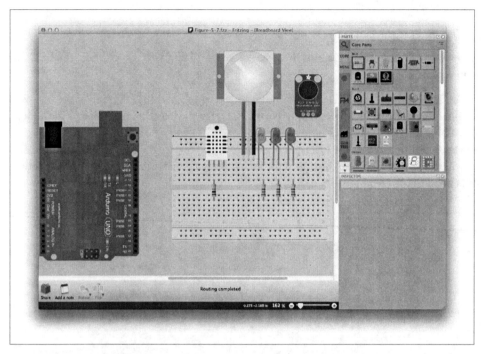

Figure 5-7. Fritzing Diagram with parts placed but unwired

Once we have placed our components, we need to make the appropriate electrical connections by adding wires between pins on the Arduino and locations on the breadboard corresponding to the components we've placed.

We can easily draw wires by clicking and dragging off of any place a connection can be made. For our circuit, this includes the Arduino header, the holes on the breadboard, or the legs of a component like the LEDs. Be sure to connect each end of the wire to a component in the circuit.

Wires can terminate on components, but wires cannot terminate on other wires.

You will notice a red dot at the unconnected end of a wire, and a green dot at the connected end of a wire. As you drag the ends of your wires around to make connections, anything currently connected to the wire end being dragged will turn yellow to clearly highlight existing connections.

Go ahead and drag a wire from the GND pin on the Arduino to the hole on the lower-right corner of the breadboard adjacent to the blue line. When you are done, your circuit should look like Figure 5-8.

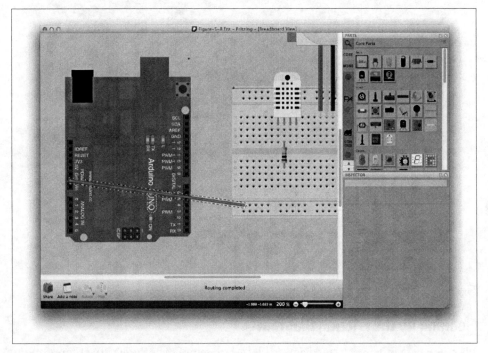

Figure 5-8. Drawing our first wire

Repeat this process for the other wires in our circuit. Remember that we are duplicating each wire in the physical circuit with a corresponding wire in Fritzing.

When you have completed this process, your circuit will look something like the circuit in Figure 5-9. It's okay if your circuit is not exactly like the Frizting diagram pictured here. What is important is that the connections you've made in your Fritzing circuit diagram accurately match the connections from your actual circuit board.

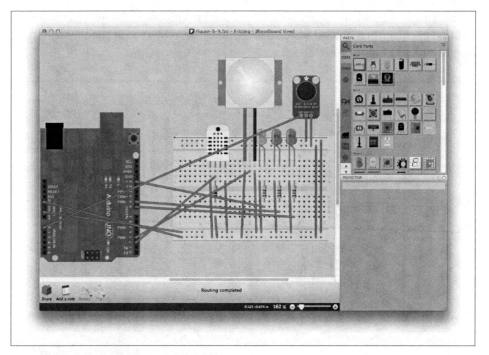

Figure 5-9. Fully wired Fritzing Circuit

Cleaning up the Fritzing Diagram

We now have a Fritzing circuit diagram that accurately depicts the electrical connections present in our actual circuit. However, before we share this diagram or use it as a guide for soldering together our circuit, it's a good idea to clean up the routing and coloring of the wires. We want the wire paths to be clear and easy to follow. We can achieve this by changing the color of the wires so that similar signals are represented by similar colors.

 As mentioned in Chapter 2, a common convention for 5V circuits is to make the +5V (VCC) wires red and the ground (GND) wires black.

Let's go ahead and change the wire colors. Right-clicking on any wire brings up a menu with "Wire Color" as an option. We can choose our colors from this menu, or we can click on a wire and select the color that we want from the "Properties" box in the lower-right corner of the Fritzing window.

Once we assign colors to our wires, our circuit will begin to be easier to understand. The color-coded version of our Fritzing diagram is shown in Figure 5-10.

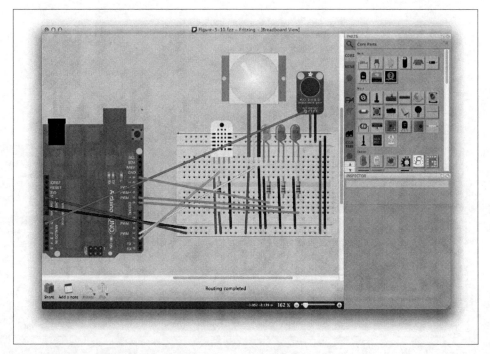

Figure 5-10. Fritzing diagram with color-coded wires

As you can see, some of the wires are crossing over other wires or circuit components. But with a little bit of work, we can reroute those wires so that all the connections and components are clearly visible.

To do this, we simply need to add bends and angles to our wires so that they route around other elements on the circuit board rather than over or through those objects. To create a bend at a given point along a wire, just click that point and drag it to a new location. You now have a bend point.

 It is a common practice to route wires either vertically or horizontally, and to replace right-angle bends with a 45° bevel. However, this is not a hard and fast rule; our primary interest is to make a circuit diagram that is easily readable.

When you are all done, your circuit should look similar to the Fritzing diagram in Figure 5-11.

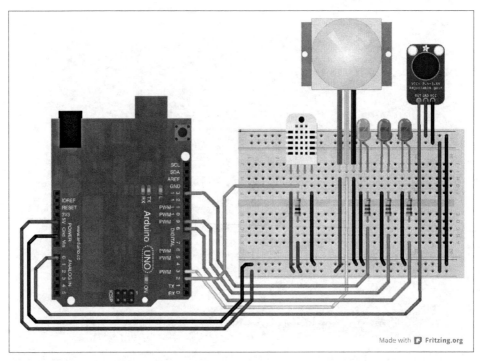

Figure 5-11. Final Fritzing diagram

We are now finished documenting our project in preparation for building an actual circuit board. Clear documentation of our board will make it much easier to understand and follow the wiring of our project. We can quickly share this project with other people in the form of a Fritzing file or export it as a picture. We can also rapidly convert the Fritzing circuit diagram into a printed circuit board (PCB) diagram in order to send our project to a factory where it can be manufactured by the dozens or by the thousands.

 If you want to learn more about making Printed Circuit Boards from your Fritzing diagrams, there is a short tutorial on the Fritzing website (*http://fritzing.org/learning/tutorials/designing-pcb/*).

Now we are ready to move our physical circuit components off of the breadboard, and connect them together using a more permanent, secure method.

Dropping Breadboard Crumbs

With our breadboard safely documented in Fritzing, we can begin to move our parts from the breadboard to a more permanent location. The primary benefit of solderless breadboards is that our parts are only held in place by friction. This is also the primary weakness of solderless breadboards.

If you were to hold your breadboard upside down and shake it, your Arduino would become disconnected and your motion sensor would fall out. Now you might be thinking: *Who puts a breadboard upside down and shakes it?* Truthfully, that exact scenario might be rare. However, you will find is that it doesn't take much to dislodge a wire and cause your circuit to stop working. A little bump or a light tug is all that it takes to cause wires to become disconnected.

Sometimes you won't notice that a wire has come loose, and you may spend hours trying to figure out what is wrong with your circuit. It's also not unheard of for a dislodged wire to cause a small breadboard fire. We need a better solution than wires held in place by friction if we want to send our circuit out into the real world.

Making the Board Permanent

Fortunately, we can make secure electrical and mechanical connections using solder, the duct tape of the electronics world. To get started, we will need some tools and materials, including a spool of solder, a soldering iron, and a version of our breadboard that we can solder parts to, which is known as a *protoboard*.

Solder

Solder is a metal alloy with a low melting point generally consisting of a mixture of tin and lead, or tin, silver, and copper (see Figure 5-12). Solder can be found at most stores that sell electronics components, like Radio Shack.

 You should not use solder from the plumbing section of hardware stores. Plumbing solder generally melts at a high temperature that would damage electronics.

Figure 5-12. Basic lead-free solder

 Make sure your solder contains a rosin core flux. This compound helps the solder more uniformly coat the wires or component leads in order to form a stronger electrical and mechanical bond. Most electronics solder contain a rosin core flux.

Soldering Irons

To heat up our wires and pins so they can be soldered, we will use a soldering iron. There are a wide variety of soldering irons available from common electronics suppliers. All of them feature a pencil-like tip that heats up above the melting point of the solder we will be using. Some of the more expensive models have useful features like a temperature control knob (see Figure 5-13).

For most electronics work, a basic soldering iron is sufficient. You can purchase them from a wide variety of vendors, including the Maker Shed, Adafruit, Sparkfun, and Radio Shack.

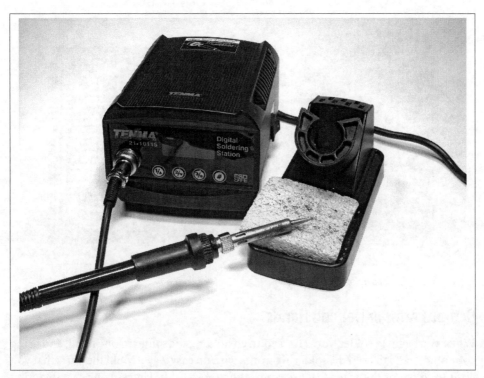

Figure 5-13. Temperature-controlled soldering iron

The Protoboard

Protoboards are printed circuit boards with copper-plated hole patterns and copper trace patterns that make it easy to solder down components and create electrical connections. Some protoboards are designed to duplicate the holes and wire connections of the solderless breadboard that we have been using. Products like the Adafruit Perma-Proto (see Figure 5-14 or go here (*http://www.adafruit.com/products/591*)) even match the color scheme and labeling of our solderless breadboard.

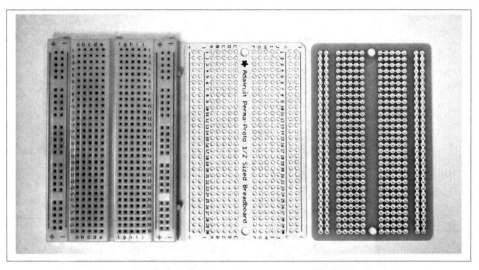

Figure 5-14. Adafruit Perma-Board

Octopus Arms or Helping Hands

When you begin to solder, you'll find yourself holding the solder in one hand, and using your other hand to hold the soldering iron—with no easy way to hold the parts that you want to solder or the wires that are connecting those parts. Unless you are a soldering octopus, you will need at least one more tool to solder successfully.

 It's generally a good idea to have a small bench vice, like a Panavise, to hold your board or part, as well as a device commonly known as *helping hands* for holding wires.

Helping hands are great for two reasons. First, as you solder wires to your project, the copper in the wire will conduct the heat from the soldering iron to your fingertips, if you are holding these wires with your fingers. Helping hands prevent your fingers from getting burned. Second, it is very difficult to hold a wire perfectly still with your fingertips while you are soldering (especially if you are burning those fingertips!). As your fingers shake and your wire wiggles, small cracks form in the solder as it cools around your wire. These cracks cause what is called a *cold solder joint*, where the bond between the wire and solder is compromised.

Let the Solder Flow

Soldering is a skill to be practiced, not an art to be mastered. There are a few basic rules to follow.

The most important rule is: don't burn yourself! Molten solder is hot, as is the soldering iron that you use to melt the solder in the first place. Also, molten solder can splatter. It's a critical practice to wear eye protection while soldering to keep hot solder droplets from splashing into your eyes.

Next, you need to know that solder flows to where the heat is. If you want to apply solder to a pin on your circuit, you must first heat the pin until it is hot enough for the solder to melt and flow onto that pin. If you are soldering a pin or wire to a pad on a circuit board, you need to heat both the pin and the pad simultaneously. This ensures that you will get solder covering both the pin and the pad, creating a strong electrical and mechanical connection.

This brings us to our next rule: don't apply too much heat! This is where soldering gets tricky. You need to apply just enough heat to melt the solder onto your part and the pad you are soldering to. If you overheat most electronic parts, you can damage them. It's also possible to overheat a pad and cause it to lift off the circuit board, rendering it useless.

Soldering is a tricky balance between applying just enough heat, and not too much heat. How do you know when you've added enough heat? Generally when the solder has melted around a pin or wire, you can remove the heat.

Let's get started soldering components to our protoboard!

The first part that we will solder to our protoboard is a resistor. It is less prone to being damaged by heat, and we can easily bend the legs to help hold it in place while we work. Let's solder a 10kΩ resistor to the left side of our protoboard, as shown in our Fritzing diagram.

To solder the resistor, first bend the legs at a right angle to the body of the resistor. Next, insert them through the appropriate pads (see Figure 5-15). Make sure you are matching the Fritzing diagram.

After you've inserted the leg of the resistor through the pad on the protoboard, you might find it helpful to give the leg a slight bend to help hold it in place while your solder it down.

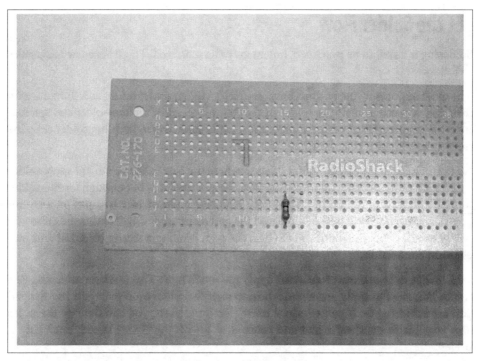

Figure 5-15. Protoboard with resistor ready to solder

Now comes the fun part. Turn your soldering iron on. Make sure the temperature is set somewhere around 700°F if you are using a temperature-controlled soldering iron. Test the tip to make sure it is hot enough by rubbing a bit of your solder along the tip (see Figure 5-16). When the solder quickly melts to the tip of your iron, you are ready to go.

You may want to clean the tip before the next step. Dragging the tip of your iron across a wet sponge will remove any dirt and excess solder, and leave the tip of your iron shiny and perfect for transferring heat effectively.

Now place the tip of the iron so it touches both the leg of the resistor and the pad of the protoboard. Feed your solder into the intersection of the resistor, pad, and soldering iron tip (see Figure 5-17).

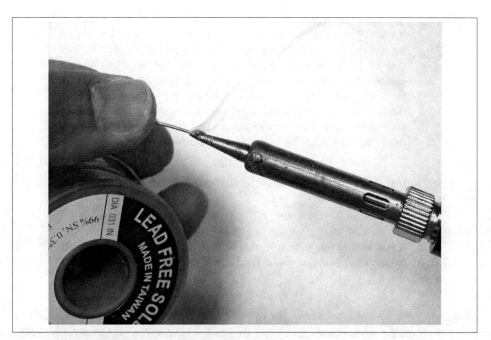

Figure 5-16. Testing heat on the soldering iron

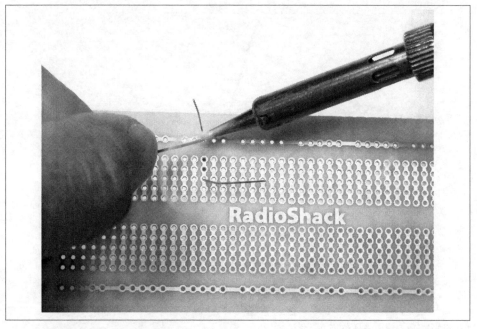

Figure 5-17. Feeding solder into the joint

If solder flows onto the soldering iron tip but not on the pad or resistor leg, you didn't transfer enough heat to the resistor leg or pad. Withdraw your solder and rotate the soldering iron tip so that the solder you just coated the tip with now covers the resistor leg and pad. This helps with the heat transfer. Try feeding your solder into the junction again. With the extra heat you've just added, the solder should now flow onto both the resistor leg and the pad, creating a tent-like shape as the solder is drawn up the resistor leg through surface tension.

Congratulations! You've just soldered your first connection. Now repeat this for the other leg of the resistor.

 If you want to take your soldering further, check out the book *Learn to Solder: Tools and Techniques for Assembling Electronics* (O'Reilly).

After you have finished soldering the resistor (see Figure 5-18), you can repeat this process for all the components and wires on the Fritzing diagram. Once you've completed soldering your components to your protoboard, it is time to run a quick test on the circuit to make sure everything works!

Figure 5-18. A properly soldered resistor

See "Communicating with Python" on page 50 in Chapter 4 if you're having trouble remembering how to test the circuit.

Summary

This chapter introduced the steps required to document your electronics project and solder your parts to a protoboard. In the next chapter we'll improve the construction of our prototype by replacing the stock protoboard with an Arduino Wireless Proto Shield.

CHAPTER 6
Simplifying the Design

In the last chapter, we documented our project with Fritzing, then moved it from the breadboard onto a protoboard, learning basic soldering skills in the process. The protoboard we used matched the form factor and footprint of our solderless breadboard, making the transfer process easy with a one-to-one match between the pads and traces on the protoboard and the connections present in the solderless breadboard.

We now have a prototype that matches our Fritzing diagram. This prototype solves some of our problems. (We no longer have a bunch of parts that can easily become dislodged from our solderless breadboard.) We can still make this prototype better, though.

It would be nice if we could eliminate the wires connecting the two circuit boards by using a protoboard that plugs directly into the Arduino. It would also be nice if that circuit board had a socket where we could plug in an XBee wireless module.

Fortunately, a solution exists that achieves all of these goals, and makes our prototype cleaner and more simple in the process.

Arduino Proto Shields

For some simple projects, the basic protoboard might be sufficient to create a nicely designed and rugged prototype. But for many Arduino projects, the ideal protoboard would have just the right pins and holes to plug into an Arduino. Fortunately, such boards are widely available and very affordable. These boards are called Arduino Proto Shields (see Figure 6-1).

Figure 6-1. Arduino Proto Shield

They are designed so that we can easily build Arduino-based prototypes with wired-up switches, sensors, motors, wireless transceivers, or other devices. These boards contain a prototyping grid—essentially just a rectangular grid of conductive pads that are plated through their thickness with metal for ease of soldering. We can solder components to this prototyping grid, and very easily connect those components to the various signals on the Arduino.

There are a few notable differences between these Arduino Proto Shields and the protoboards that we worked with in the last chapter. Unlike the matching printed circuit boards, where the pads have been connected to match our solderless breadboard, most of the pads on the Arduino Proto Shield are not connected to each other. We must solder wires between our parts to create connections.

The Arduino pins are one exception to this. The pads lying adjacent to the tall black pin connectors on the prototyping shield have labels that correspond to the names of the Arduino pins. These signals all connect directly to the matching pins on the Arduino. If we solder a wire into one of these pads, that wire will carry a signal to or from the corresponding Arduino pin.

Each Arduino signal is passed up through the black stacking headers. This allows us to plug multiple shields into the same board, in case we can't fit all our parts on one prototyping shield.

Arduino prototyping shields are manufactured by a number of different vendors. There are special variants of these shields with extra features, like memory card connections or XBee wireless interfaces. We will be working with a variant of the prototyping shield that has a built-in connector for wireless devices like the XBee (see Figure 6-2).

Figure 6-2. Arduino Wireless Proto Shield with XBee

Fritzing Revisited

Let's take a look at our Fritzing circuit again (see Figure 6-3). We used a solderless breadboard in our Fritzing diagram because it matches exactly the physical circuit that we had built.

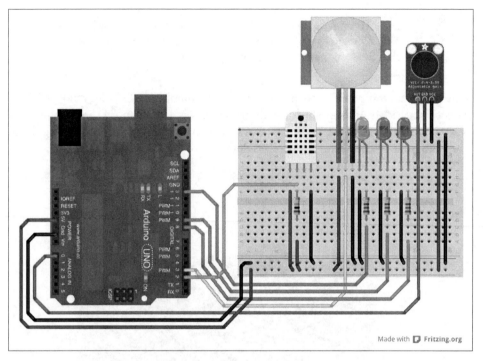

Figure 6-3. Our original Fritzing circuit

We now want to switch over from a solderless breadboard to an Arduino Protot Shield. We can see what this will look like in Fritzing before we make any changes to our actual circuit.

Begin by selecting the Arduino parts bin. There is a part called "Arduino Wireless SD Shield" in the Arduino parts bin. Select this part and drag it anywhere in your Fritzing diagram (see Figure 6-4). This part has connection points that exactly match the Arduino that you already have in your Fritzing diagram.

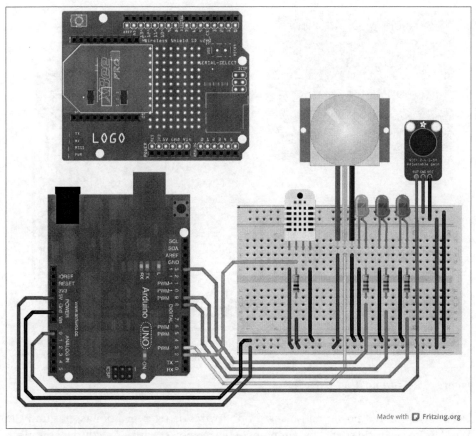

Figure 6-4. Arduino and Wireless Proto Shield in Fritzing

Since the connections on both the Arduino and the Wireless Proto Shield match, we can delete the Arduino from the Fritzing diagram, and simply replace it with the Wireless Proto Shield. Make sure to unlock the Arduino if you had locked it down, using the "locked" checkbox in the Parts Inspector. Then click on the Arduino and hit Delete. All of the wires that were connected to the Arduino are now floating at the ends (see Figure 6-5).

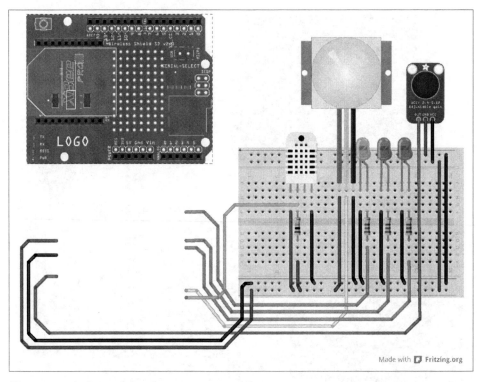

Figure 6-5. Arduino deleted

Now, using the red power and black ground wires as reference points, position the Wireless Proto Shield so that the endpoint of the red power wire is exactly over the black header labeled "5V", and the black ground wire is positioned exactly over the black header labeled "GND."

The tips of all the wires won't automatically snap onto the Wireless Proto Shield. You will need to click and drag the wires into the correct location.

 You may have to turn off the "Align to Grid" feature if you are having trouble getting the wires to snap onto the Proto Shield. If that is the case, go to the "View" menu and uncheck "Align to Grid."

Once you have successfully replaced the Arduino with the Wireless Proto Shield, your Fritzing diagram should look like Figure 6-6.

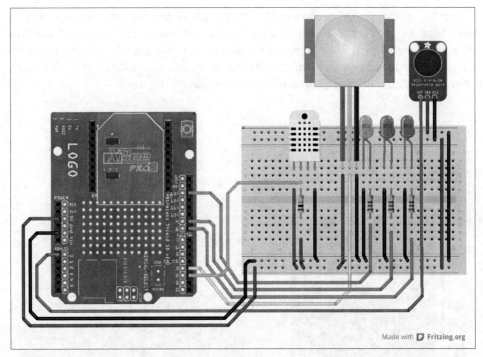

Figure 6-6. Fritzing diagram with only the Wireless Proto Shield

Go ahead and lock the Wireless Proto Shield and turn "Align to Grid" back on if you had turned it off. We have taken the easy initial step of moving all our parts onto the Wireless Proto Shield by using it to replace the Arduino in our circuit. Now we have to move all the components and their associated wires off the breadboard and onto the Proto Shield.

 Fritzing, like many programs, puts the most recent parts in "Front" of the other parts. Since the Wireless Proto Shield is the last part we placed on the board, it will obscure all the other parts. Right click the Proto Shield and select "Raise and Lower -> Send to Back" so you can see your parts as you move them onto the Proto Shield.

If you try to move your parts to the prototyping area of the Wireless Proto Shield, several problems will quickly become apparent. First, there is hardly enough room to squeeze in all of your parts. Second, we don't have enough power or ground connections on the Wireless Proto Shield to connect to our parts.

When we were working with the solderless breadboard or the matching protoboard, we had convenient power and ground busses to which we could wire all of our parts, and each of those busses could be connected to the Arduino using a single power and ground wire. Now we need six grounding points, but we only have three. We would also like more than one 5V pad so we can power the motion detector, the microphone, the DHT 22 humidity and temperature sensor, and the pull-up resistor for the DHT 22. We won't get very far with our Fritzing diagram before it looks like Figure 6-7.

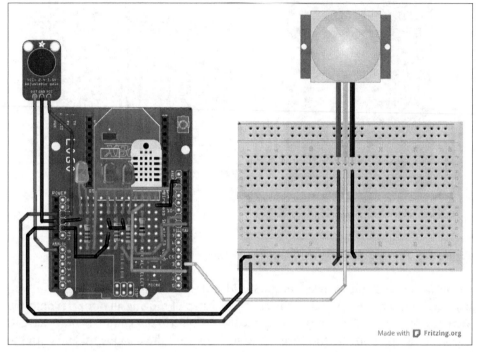

Figure 6-7. Fritzing diagram: moving components onto the Wireless Proto Shield

We do have a simple solution to this problem. It is a combination of a clever hardware hack plus a little software support.

Power and Ground

Every component in our circuit has at least two pins. Electricity flows into one of the pins, and flows out the other pin. To use a standard convention, the pin into which positive electrical charge flows is labeled "positive" or "power," and the pin that this charge flows out of is labeled "negative" or "ground." Our LEDs light up when positive charge flows into the positive leg (the *anode*) and flows out of the negative leg (the *cathode*). See Figure 6-8.

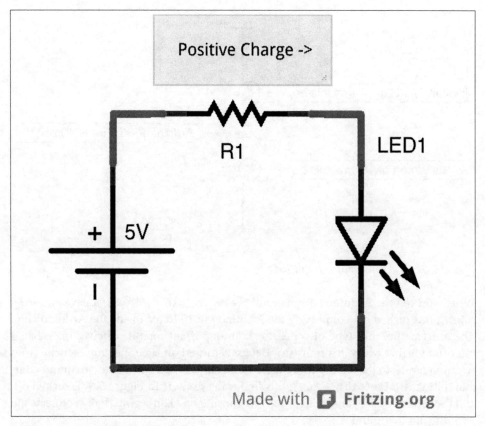

Figure 6-8. Simple LED circuit

Similarly, our microphone is active when its power lead is connected to a 5V power source and the ground lead is connected to a point that provides ground. It may seem like there are only a few locations on our Arduino Proto Shield that give us power and ground connections. But as it turns out, there are a lot more power and ground connections than we may think. Let's take a look at a basic Arduino pin (Figure 6-9).

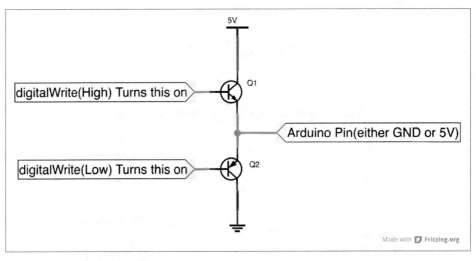

Figure 6-9. An Arduino pin simplified

When you call the Arduino command `digitalWrite(pin10, High)`, a device called a *transistor* is turned on, connecting the Arduino pin 10 to 5V inside the Arduino chip. The exact nature and type of transistor is unimportant for this discussion. What is important is that when this transistor device is turned on, a connection is made from Arduino pin 10 to the 5V input to the Arduino chip. When you call the Arduino command `digitalWrite(pin10, Low)`, this first transistor, Q1 in Figure 6-9, is turned off, and the second transistor, Q2 is turned on. Turning on Q2 now internally connects the Arduino pin to ground.

 Q1 and Q2 will never be on at the same time. They may both be off if our Arduino pin is either not initialized, or is initialized as an "input" using the `pinMode(pin10, Input)` command.

What is very useful about this design of the Arduino pins is that we can use them to provide power and ground to other parts of our circuit. If we need to power a device, we simply plug that device into an Arduino pin, set the pin to be an output, and set the output level to be **High**. Similarly, we can provide a ground line by setting a pin to be an output, and setting the output level to be **Low**.

In our Arduino code, we need to configure all of the lines that will be used as power or ground, in addition to configuring the lines that will be used to read sensors or drive LEDs:

```
#include <DHT.h>

#define DHTTYPE DHT22      // DHT 22  (AM2302)
#define SILENT_VALUE 380   // Starting neutral microphone value (self-correcting)

// LEDs
int statusLED = 13;        // The status LED uses the adjacent ground pin

int motionLED = 2;
int motionLED_gnd = 3;

const int powerLED = A5;
const int powerLED_gnd = A4;

// DHT
int dhtPin_5v = 11;
int dhtPin = 10;
int dhtPin_gnd = 8;
DHT dht(dhtPin, DHTTYPE);

// PIR
int pirPin_gnd = 5;   ❶
int pirPin = 6;
int pirPin_5v = 7;    ❷

int pirState = LOW;        // Start off assuming no motion detected
int pirVal = 0;
int motionState = 0;

// MIC
const int micPin = A0;
const int micPin_gnd = A1;
const int micPin_5v = A2;

int micVal = 0;

// Setup
void setup() {

  // LEDs
  pinMode( statusLED, OUTPUT );
  pinMode( motionLED, OUTPUT );
  pinMode( powerLED, OUTPUT );

  digitalWrite(powerLED, HIGH);   // Power ON

  // DHT
  pinMode( dhtPin, INPUT );

  // PIR
  pinMode( pirPin, INPUT );
```

```
    pinMode( dhtPin_5v, OUTPUT); ❸
    pinMode( pirPin_5v, OUTPUT );
    pinMode( micPin_5v, OUTPUT );

    digitalWrite( dhtPin_5v, HIGH );  ❹
    digitalWrite( pirPin_5v, HIGH );
    digitalWrite( micPin_5v, HIGH );

    pinMode( motionLED_gnd, OUTPUT ); ❺
    pinMode( powerLED_gnd, OUTPUT );

    pinMode( dhtPin_gnd, OUTPUT );
    pinMode( pirPin_gnd, OUTPUT );
    pinMode( micPin_gnd, OUTPUT);

    digitalWrite( motionLED_gnd, LOW ); ❻
    digitalWrite( powerLED_gnd, LOW );
    digitalWrite( dhtPin_gnd, LOW );
    digitalWrite( pirPin_gnd, LOW );
    digitalWrite( micPin_gnd, LOW );

    // Begin
    Serial.begin(9600);
    dht.begin();

}

void loop() {
    digitalWrite(statusLED, HIGH);

    // DHT-22
    float h = dht.readHumidity();
    float t = dht.readTemperature();

    // PIR
    pirVal = digitalRead(pirPin);   // read input value
    if(pirVal == HIGH){             // check if the input is HIGH
      if(pirState == LOW){
        // we have just turned on
        motionState = 1;
        digitalWrite(motionLED, HIGH);  // turn LED ON

        // We only want to print on the output change, not state
        pirState = HIGH;
      }
    } else {
      if(pirState == HIGH){
        // we have just turned of
        motionState = -1;
        digitalWrite(motionLED, LOW);// turn LED OFF

        // We only want to print on the output change, not state
```

```
      pirState = LOW;
    }
  }

  // Microphone
  micVal = getSound();

  // Check that we have a reading from the DHT sensor
  if (isnan(t) || isnan(h)) {
    Serial.println("Error: Failed to read from DHT");

    Serial.print(pirVal);
    Serial.print(",");
    Serial.print(motionState);
    Serial.print(",");
    Serial.println(micVal);

  } else {
    Serial.print(h);
    Serial.print(",");
    Serial.print(t);
    Serial.print(",");
    Serial.print(pirVal);
    Serial.print(",");
    Serial.print(motionState);
    Serial.print(",");
    Serial.println(micVal);

  }

  // reset motion state
  motionState = 0;

  // end packet
  digitalWrite(statusLED, LOW);
  delay(2000);

}

int getSound() { ❼
  static int average = SILENT_VALUE;
  static int avgEnvelope = 0;
  int avgSmoothing = 10;
  int envSmoothing = 2;
  int numSamples=1000;
  int envelope=0;
  for (int i=0; i<numSamples; i++) {
    int sound=analogRead(micPin);
    int sampleEnvelope = abs(sound - average);
    envelope = (sampleEnvelope+envelope)/2;
    avgEnvelope = (envSmoothing * avgEnvelope + sampleEnvelope) /
                  (envSmoothing + 1);
```

```
        average = (avgSmoothing * average + sound) / (avgSmoothing + 1);
    }
    return envelope;
```

❶ Here, we set pin 5 to be the GND line for the PIR sensor.

❷ This line sets pin 7 as the 5 volt line for the PIR sensor.

❸ Set these pins to be outputs that will supply 5V to the sensors.

❹ Here, we turn on 5V for the sensors.

❺ Set these pins to be outputs that will supply GND for the sensors.

❻ Here, we turn on GND for our sensors.

❼ See the end of "Modifying the Software" on page 38 (after the code block) for an explanation of this code.

Cleaning up the Fritzing Diagram

Now that we have a way to get power and ground from all the pins on our Arduino, moving the components in our Fritzing diagram from the solderless breadboard over to the Arduino Wireless Proto Shield should become much easier. We do not have to run separate power and ground lines to every part. We can plug those parts directly into the Arduino pins.

You will find it very difficult in the Fritzing Breadboard View to create wires that go from the end of one part—say, an LED—to the end of another part, like a resistor. The breadboard view generally requires you to plug parts into a breadboard or a similar base like an Arduino or Arduino Shield. In order to make connections between the LEDs and resistors, as we have, you will need to switch to the Fritzing Schematic View, where you should be able to easily make those connections. After you make a connection in the Schematic View, switch back to the Breadboard View. You will now see a thin wire representing the connection that you just made. Right click on this wire and select "Create Wire From Ratsnest." You will then have a wire in your Breadboard View.

With a little bit of quick work, you should have a Fritzing diagram that looks clean and simple, with the minimal number of wires used to connect all of your components to the Arduino Wireless Proto Shield (see Figure 6-10).

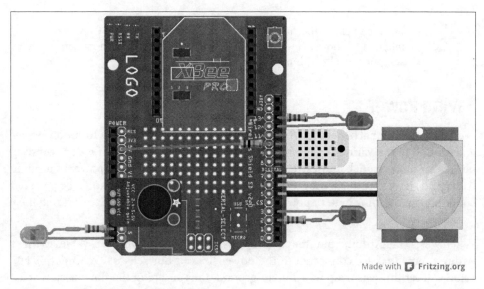

Figure 6-10. The simplified Fritzing diagram

Rules for Pin Power and Ground

In many instances, this trick of using the Arduino pins themselves to supply power and ground is very convenient. But this hack will not work in every scenario. The Arduino DigitalPins page (*http://arduino.cc/en/Tutorial/DigitalPins*) explicitly warns us about drawing too much power from these pins:

> Short circuits on Arduino pins, or attempting to run high current devices from them, can damage or destroy the output transistors in the pin, or damage the entire Atmega chip. Often this will result in a "dead" pin in the microcontroller but the remaining chip will still function adequately.

How much current is too much current? Looking at the Arduino hardware pages (*http://arduino.cc/en/Main/Products?from=Main.Hardware*), we find our answer:

> Each of the 20 digital i/o pins on the Leonardo can be used as an input or output, using pinMode(), digitalWrite(), and digitalRead() functions. They operate at 5 volts. Each pin can provide or receive a maximum of 40 mA and has an internal pull-up resistor (disconnected by default) of 20–50 kOhms.

As long as we keep the current into or out of the I/O lines below 40 mA, we will not cause problems for the Arduino.

Be sure to check the current consumption of any device that you are intending to power from Arduino pins to make sure the current consumption will be within the 40mA limitation.

Saving Power

We get one additional benefit from using the Arduino pins themselves to supply power and ground to the various components we are using in our circuit. If we are running our sensor motes from batteries, it is important to be able to conserve power to prolong battery life. If we wire our sensors directly to power and ground, they would be powered continuously, draining battery power regardless of whether or not we are reading their values.

When we use the Arduino pins for power, we can be clever and write **LOW** values to those Arduino pins that are connected to the power lines of our sensors. Writing this **LOW** value will set the sensor's power line to ground, thus stopping the flow of electricity. When we are ready to read the sensor, all we need to do is set the Arduino pin corresponding to the sensor power line to **HIGH**. This will restore the flow of electricity to the sensor, and allow us to take our readings.

Summary

In this chapter we made a significant improvement in the construction of our Arduino prototype by replacing the stock protoboard with an Arduino Wireless Proto Shield. We also simplified our circuit by taking advantage of the Arduino pins as sources of power and ground for our circuit components. This simplification paid dividends with power consumption.

CHAPTER 7
Building Point-to-Point XBee Networks

Up until now, our Arduino has been connected to our computer via a USB cable. In order to use our Arduino and the attached sensors in the environment, we need to eliminate the USB cable and replace it with a wireless connection.

In the past, adding wireless communication to a project required extensive electrical engineering expertise. The available off-the-shelf wireless modules (Figure 7-1) were typically bulky proprietary technologies that required customization and debugging to use with a particular application.

Today, we have many compact, easy-to-use modular wireless solutions available to us. These modules use commonplace standards like Wi-Fi, GSM/CDMA (cellular telephone), Zigbee, and Bluetooth. Each one of these standards has tradeoffs that make them more or less appropriate for a particular application.

One of the standards used most commonly when dealing with sensor networks is IEEE 802.15.4. This protocol focuses on low-cost, low-speed communication between devices and is intended for low-power scenarios. It is the basis for the higher-level ZigBee protocol (also commonly used when dealing with sensor network), which further extends the standard by developing the upper layer protocols to enable mesh networking between devices.

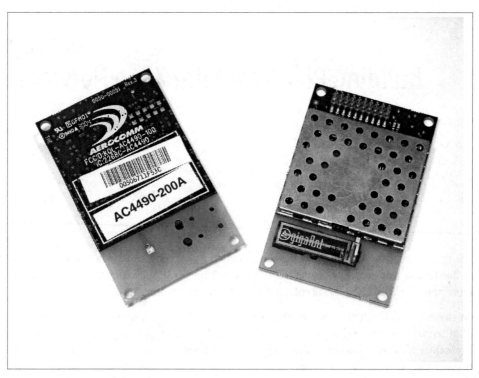

Figure 7-1. Aerocomm module

While many other brands of 802.15.4-based hardware exist, probably the most widely used are the XBee branded radios. They are used by professionals during prototyping, but also in the hobbyist market because they're especially easy for beginners to use. For this and other reasons we will focus on the XBee-branded radio modules are manufactured by Digi International (*http://www.digi.com/*).

XBee Modules

Digi manufacturers a confusingly large array of XBee-branded radios: a quick count shows that there are at least 30 different combinations of hardware, firmware, and antenna options, and this number is only going to get larger. Unfortunately, it's very easy to get confused by the growing range of XBee-branded radio modules.

Series 1 or Series 2?

There are two types of pin-compatible XBee modules: Series 1 and Series 2. One of the misunderstandings when dealing with XBee modules is that somehow Series 2 modules are "better" than Series 1 modules. That's just not true. Instead, the two types are intended for different purposes. The Series 1 module is intended for point-to-point and point-to-multipoint applications, while the Series 2 module is intended for applications that require mesh networking.

The Series 1 modules are popular with the hobbyist market, and are an excellent fit for a straightforward cable replacement, or for small networks. Series 2 implements the full Zigbee protocol, and are intended for larger networks.

While the XBee Series 1 and the XBee Series 2 modules have the exact same form factor and are pin-for-pin compatible, they are based on different chip sets and are running different protocols, so they are not over-the-air compatible. You cannot mix Series 1 and Series 2 modules in the same wireless network.

A Series 1 module will work out of the box for point-to-point communication (i.e., cable replacement) without any work, which is how we're going to be using them in this chapter. So if you're just starting out with wireless networks, start with the Series 1 radios.

If you're interested in building larger XBee-based networks, you might want to look at *Building Wireless Sensor Networks* by Robert Faludi (O'Reilly). While that book uses Series 2 modules exclusively, it can also help you with Series 1 networks, as Series 1 commands are, for the most part, just a subset of the Series 2 command set. If you are interested in other interesting projects that make use of the Series 1 modules covered in this chapter, check out *Making Things Talk* by Tom Igoe (O'Reilly).

We are using Series 1 radios (shown in Figure 7-2) mainly because they're very easy for beginners to use and to configure, and, perhaps more importantly, easy to obtain. You can get the Series 1 radios from a number of vendors, including SparkFun, who carry a wide range of modules (*http://www.sparkfun.com/categories/111*). However, despite the dizzying range on display, they still don't carry them all. You can also get Series 1 radios at Maker Shed (*http://www.makershed.com/ProductDetails.asp?ProductCode=MKAD14*).

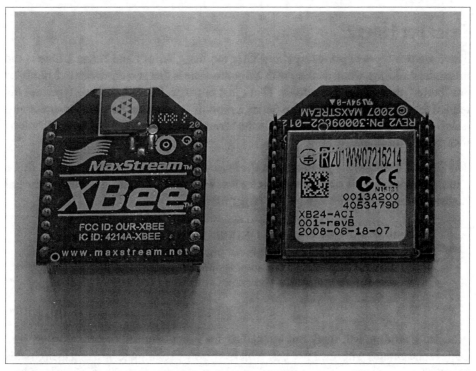

Figure 7-2. An XBee Series 1 module with a chip antenna

Regular vs Pro?

Both Series 1 and Series 2 modules come in two flavors; regular and Pro. The Pro versions of the modules are slightly larger, although despite this they are still pin-compatible, and use more power. Crucially, however, they have a much longer range. Predictably, they also cost a lot more, so unless you need the extended range they offer, you should stick with the regular XBee modules. (Although, since they're pin-compatible, you can always drop a Pro into your build later on if you decide you need the extended range, and can afford the increased power load.)

Table 7-1. XBee regular vs. Pro comparison

XBee Module Type	Regular	Pro
Indoor Range	100 ft (30 m)	300 ft (90 m)
Outdoor Range (Line of Sight)	300 ft (100 m)	1 mile (1.6 km)
Transmit Current	45 mA	215 mA
Receive Current	50 mA	55 mA

802.15.4 or Zigbee?

There is a lot of confusion surrounding the two XBee protocols. Series 1 modules use the IEEE 802.15.4 standard protocol that allows point-to-point and point-to-multipoint networking, although they also have a proprietary mesh networking protocol. Series 2 modules use the Zigbee protocol, which is a mesh-networking standard built on top of the 802.15.4 protocol that is spoken by numerous other pieces of hardware, not just the XBee radio modules.

Which Antenna?

There are four different types of antennas on offer: chip, wire, UFL, and RP-SMA. The chip and wire antennas come pre-connected to the radio module, however the UFL and RP-SMA based offerings ship with connectors only on the board. You'll need to purchase an appropriate antenna with the proper connector to get them to function. Unless you're intending to enclose your project in a box—at which point mounting the antenna outside the box using a UFL or RP-SMA antenna is probably a good idea—or need the extended transmission distance they provide, you should be safe enough sticking with a chip- or wire-based antenna.

How to Configure an XBee Series 1 Radio

To configure an XBee module, we need to be able to talk to it. Generally, we'll be using your Mac to configure your XBee, so we'll need a way to connect the XBee to your Mac's USB port.

Since they were designed to be soldered directly to a PCB (printed circuit board), XBee modules rather unfortunately use 2mm headers instead of the more familiar 0.1-inch headers that we are all so used to. They generally therefore require a breakout board of some kind before they can be used on a standard bread- or protoboard, and a USB adaptor before they can be connected to your Mac.

There are a number of breakout boards on the market; generally, you'll just need a simple board that breaks the XBee pins out to standard 0.1-inch header blocks, like SparkFun's Breakout Board for XBee Module (*http://www.sparkfun.com/products/8276*).

To connect the XBee to your computer, you'll need a more complicated board that takes care of regulating the input voltage to the 3.3V required by the module; signal conditioning; and providing basic activity-indicator LEDs. Again, there are a large number of these on the market, like SparkFun's XBee Explorer Regulated (*http://www.sparkfun.com/products/9132*)), or Adafruit's XBee Adaptor Kit (*http://www.adafruit.com/products/126*). See Figure 7-3.

There are even boards that allow you to plug the XBee directly into your Mac's USB port, such as SparkFun's XBee Explorer Dongle (*http://www.sparkfun.com/products/9819*) (see Figure 7-3 again).

 Almost invariably, XBee USB adaptors require drivers from FTDI (*http://www.ftdichip.com/Drivers/VCP.htm*). Make sure you have installed these before using your adaptor. If you have an older Arduino board, you may have already installed the correct driver, as boards before the Uno made use of the same USB chipset.

Figure 7-3. The Adafruit XBee Adaptor (left) and the SparkFun XBee Explorer (right)

Connecting the XBee to your Mac

We are going to use the Adafruit XBee Adapter board to get started. It's designed to be used with an FTDI cable (see Figure 7-4), and because of that it can also be easily used with a breadboard to connect the XBee to an Arduino, so we won't need a separate breakout board.

While Digi's own configuration tool, X-CTU, is available for free, it's also Windows-only. You can download X-CTU from Digi's website (*http://www.digi.com/support/productdetail?pid=3352*); be sure to also check out Digi's short instructions for using X-CTU (*http://bit.ly/15zSTAj*). Unless you're running Windows under BootCamp or in a virtual machine, you won't be able to make use of it. Luckily, it's only needed infrequently to update the firmware of the XBee modules. For configuring the radios, we can generally just use a Terminal program.

Figure 7-4. An XBee module mounted on an Adafruit Adaptor Kit, connected to an FTDI cable for programming

 There are many Terminal programs available for Mac OS X. We're going to use CoolTerm by Roger Meier (*http://freeware.the-meiers.org/*). This is a relatively simple application that's perfectly suited to configuring XBee radios.

Plug your XBee radio module into your adaptor board, and connect the board to your Mac. If you're using the Adafruit board as shown, you should see the green (ASC) LED on the board start to flash.

Open the Terminal application and make sure it's set to the correct serial port. You can find out which serial ports are available by using the command line:

```
% ls /dev/tty.*
crw-rw-rw- 1 root wheel 11, 2 8 Jul 09:01 /dev/tty.Bluetooth-Modem
crw-rw-rw- 1 root wheel 11, 0 8 Jul 09:01 /dev/tty.Bluetooth-PDA-Sync
crw-rw-rw- 1 root wheel 11, 66 3 Aug 14:15 /dev/tty.usbserial-FTE4XVKD
%
```

The correct port for the XBee module should (at least normally) be fairly obvious. In our case, it is /dev/tty.usbserial-FTE4XVKD. If you are using CoolTerm, you can change the attached serial port by clicking on the Options button in the toolbar (see Figure 7-5).

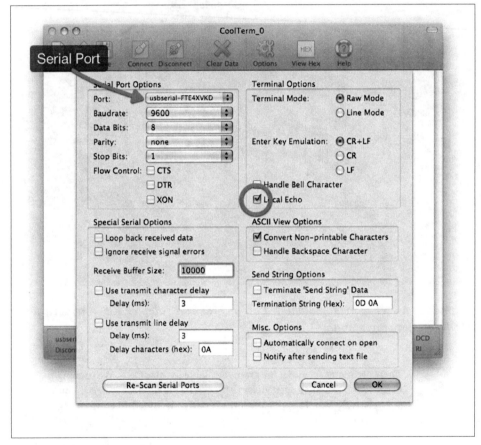

Figure 7-5. Configuring the port and local echo in CoolTerm options

By default, CoolTerm has Local Echo turned off; you may want to enable this in the Options menu to make things easier on yourself. If it's turned off, then you won't see what you're typing, and only the responses from the XBee radio will be displayed.

If your Terminal program doesn't default to it, set it to use 9,600 baud and 8-N-1 with no flow control.

Once you're sure you have your Terminal program set up correctly, connect it to the selected serial port. In CoolTerm, you do this by clicking on the Connect button in the toolbar. At this point, type **+++** (three plus signs in fairly quick succession) into the main window, if everything is connected up correctly you should get an **OK** back from the XBee radio module (see Figure 7-6).

Figure 7-6. The XBee radio module is correctly connected to your Mac

By typing the three plus symbols, we have put the XBee into configuration mode. At this point, we can send *AT* commands to the module, which will allow us to configure the radio. After a while, the XBee will time out and go back to pass-through connection mode. If that happens while you're configuring it, just type in **+++** and it will start responding once again.

The basic configuration on the XBee consists of: the baud rate for serial communication, a network identifier (PAN), the node address (MY), and the destination node address (DL).

XBee Addressing

The 802.15.4 protocol uses addressing to distinguish one radio from the next, and to prevent duplicate packets. It is very important that each module have a unique source address (MY) to prevent non-duplicate messages being ignored as duplicates.

There are two basic forms of addressing between the modules: Broadcast and Unicast.

A Broadcast message is a message that will be received by all modules on any given PAN. The message is sent only once and not repeated, so there is no guarantee of any given node receiving the message. In order to send a Broadcast message, set the destination node address (DL) to `0xFFFF`. With these settings, all XBee modules within range of the broadcasting node will receive the message.

A Unicast message is more reliable. It is sent from one module to any other module based on the module's addressing. If the message is properly received, the receiving radio will send back an acknowledgment (or *ACK*).

If the transmitting module doesn't receive an ACK, it will attempt MAC level retries—three retries for every transmission, for a total of four attempts—until it receives an ACK. This greatly increases the probability of getting the data through to the destination.

In order to send a Unicast message, set the destination node address (DL) to be the node address (MY) of the node with which you wish to communicate.

Configuring Two XBee Radios

We're now going to configure two XBee radios as a straight cable replacement. In other words, the first will be configured to send messages to the second, and vice versa.

Go ahead and plug the first of the two radios into your Mac, and place it into command mode as in the previous section by sending a **+++** to the appropriate serial port from your Terminal program.

The first thing we need to do is set the baud rate. When in configuration mode, enter **ATBD**. You should get a number between 0 and 7 returned, as follows:

```
ATBD
3
```

This number indicates that the radio is set to 9,600 baud.

We'll leave our radio at 9,600 baud for now. However, if you wanted to set the radio to a different baud rate, you'd issue the command **ATDB***n*, where *n* was a value corresponding to the baud rate you wanted from Table 7-2.

Table 7-2. The value returned by the ATBD command and corresponding baud rate

ATBD Value	Baud Rate
0	1200
1	2400
2	4800
3	9600
4	19200
5	38400
6	57600
7	115200

For example, to set our radio to 57,600 baud, we'd issue the command **ATDB6**:

```
ATBD6
OK
```

We can then check the baud rate has been set by again issuing the **ATBD** command. This time, it should return a 6:

```
ATBD
6
```

Having set the baud rate, we need to go ahead and similarly set the PAN ID using the **ATID** command, the node address using the **ATMY** command, and the destination node address using the **ATDL** command. Finally, we need to commit our changes by issuing a write command, **ATWR**, to save our settings.

Therefore, for the first of our two radios, we should issue the following commands:

```
+++
ATBD 3
ATID 1111
ATMY 2345
ATDL 7890
ATWR
```

You should get an OK back from the modem in all cases. You may want to check the values along the way. See, for instance, Figure 7-7.

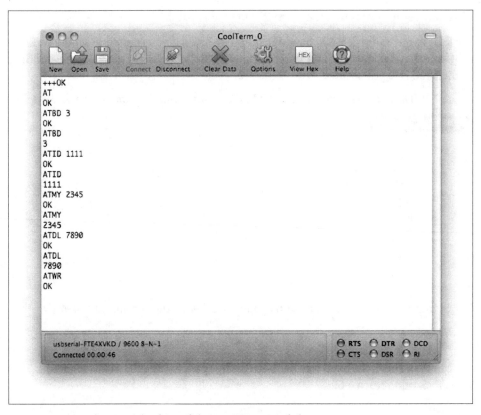

Figure 7-7. Configuring the first of the two XBee modules

Once you've configured the first radio, disconnect the serial port and unplug the XBee module from the adaptor. Then plug the second module into the adaptor, connect to the appropriate serial port, and issue the following commands:

```
+++
ATBD 3
ATID 1111
ATMY 7890
ATDL 2345
ATWR
```

This configures the second radio as the destination of the first, and vice-versa. See Table 7-3.

Table 7-3. Radio configuration

Radio	PAN	MY	DL
1	1111	2345	7890
2	1111	7890	2345

Unplug the second radio once you've finished configuring it. We'll get back to the XBee modules later. However, now that we've configured them, we're going to go ahead and test out the connection by attaching one of them to an Arduino, and the other to our Mac.

Connecting an XBee to an Arduino

You can connect and XBee radio module to an Arduino by making use of the 3.3V pin on the Arduino board to provide a regulated voltage for the module. We do need to be careful about connecting the Arduino to the XBee: while it is safe to send a 3.3V signal from the XBee to the Arduino (given that the Arduino pins can handle any signal between 0V and 5V), sending a 5V signal from the Arduino to the 3.3V XBee may cause damage. Fortunately, we only need two resistors to protect the XBee from damage, as shown in Figure 7-8.

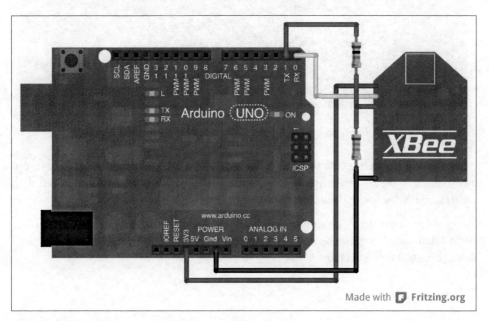

Figure 7-8. Connecting an XBee to an Arduino

It is a lot easier if you make use of a breakout board or an appropriate adaptor board, like the Adafruit board we used to connect the XBee to our Mac earlier (see Figure 7-9).

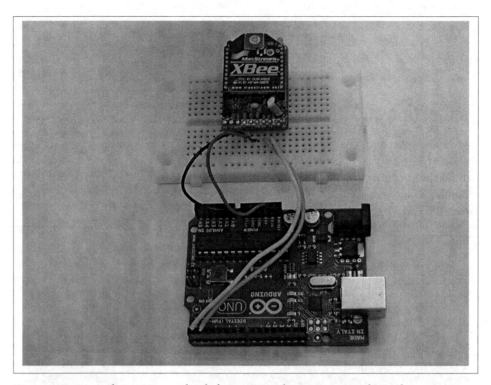

Figure 7-9. An Arduino Uno and Adafruit XBee adaptor connected together

Alternatively, if you're going to be us XBee modules a lot, you might want to invest in an Arduino Wireless Proto Shield (see Figure 7-10), which is an Arduino board with an attached XBee socket (see Chapter 6).

We're going to use the same simple sketch on our Arduino that we used in Chapter 1 to print "Hello world" to the serial port. Connect your Arduino board to your Mac, and upload the following sketch:

```
void setup() {
 Serial.begin(9600);
}

void loop() {
 while (Serial.available() <= 0) {
 Serial.println("Hello world");
 delay(300);
 }
}
```

Figure 7-10. The Arduino Wireless Proto Shield With XBee plugged in

 You cannot upload a sketch from your Mac to the board with the XBee, or any other hardware, connected to pins 0 (RX) and 1 (TX), as this interferes with the USB serial connection from your Mac. You therefore must detach the XBee from your Arduino board before uploading your sketch.

Disconnect your Arduino board from your Mac and connect it to an external power supply.

Once you've done that, attach one of the XBee radios to your Mac, open up the CoolTerm application once again, and connect to the appropriate serial port so that you can see the output from the radio.

Then go ahead and connect the other radio to your Arduino board, as in Figure 7-9 above, and power it on. You should see something very much like Figure 7-11 in your Terminal program.

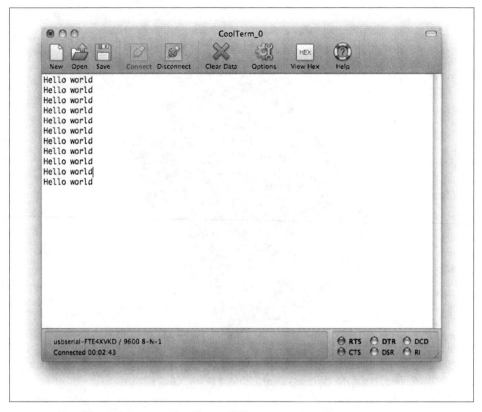

Figure 7-11. The Arduino says "Hello world" over an XBee connection

Going Wireless with XBees

A fairly trivial exercise at this point would be to take the sensor mote code and the Python example from the Chapter 5 and substitute an XBee for the direct cable connection. There aren't any software changes needed, just hardware changes. As before, you should be able to run the simple Python serial port test:

```
>>> import serial
>>> ser = serial.Serial('/dev/tty.usbmodem621')
>>> x = ser.read(30)
>>> print x
```

If you are running the demo code from the sensor mote project, the Python code should generate a printout that looks like the following:

```
22.60, 42.50, 14, 0, 0
```

Summary

In this chapter, we eliminated the USB cable on our Arduino sensor mote and replaced it with a point-to-point wireless connection using XBee radios. In the next chapter, we'll set up a many-to-point network to incorporate multiple sensor motes.

CHAPTER 8
Building Many-to-Point XBee Networks

Setting up a many-to-point or *star* network using XBee devices need not be difficult or expensive. In the previous chapter, we showed you how to set up point-to-point wireless communications, essentially replacing the USB cable that connects our Arduino to our computer with a pair of XBee modules. With a few simple steps, we can expand this point-to-point connection to incorporate multiple Arduino sensor motes.

Addressing for Multiple XBees

We can take advantage of the XBee configuration we've already walked through in Chapter 7. Remember that the basic configuration on the XBee consists of: the baud rate for serial communication, a network identifier (PAN), the node address (MY), and the destination node address (DL). The 802.15.4 protocol uses addressing to distinguish one radio from the next, and to prevent duplicate packets. It is this addressing that directs our messages to the proper location.

If we want multiple Arduino sensor motes to send data to a single XBee master device, we need to configure each of the sensor motes and the master device with the appropriate node addresses (MY) and the appropriate destination node address (DL). We will begin by assigning the node address (MY) of the XBee master device with a simple value, like 0x0001. Now the configuration of the sensor motes is fairly straightforward for communication with the master (see Table 8-1).

Table 8-1. Radio configuration

Radio	PAN	MY	DL
1	1111	0011	0001
2	1111	0012	0001
3	1111	0013	0001
4	1111	0014	0001

Notice that we are only changing the value of the node addresses (MY).

Configuration of the XBee master device is just as easy. When we were using two XBees to communicate with each other, we were using the XBees in Unicast mode. To take advantage of a simple point-to-multipoint configuration, we need to switch the master device to Broadcast mode.

Remember that a Broadcast message is a message that will be received by all modules on any given PAN. The message is sent only once and not repeated, so there is no guarantee of any given node receiving the message. This is not a problem, however. We can take advantage of the processing capabilities of the Arduino to guarantee that messages are sent and acknowledged.

In order to send a Broadcast message, set the destination node address (DL) of the XBee master device to 0xFFFF. With this setting, all XBee modules within range of the broadcasting node will receive the message. The full settings for the master device are shown in Table 8-2.

Table 8-2. Radio configuration

Radio	PAN	MY	DL
XBee master device	1111	0001	FFFF

Now, any message we send from the XBee master device will be broadcast to our sensor mote devices 1 thru 4.

This only solves half the problem. If one of our sensor motes sends data to the master device, how can we know which device transmitted the data? This is easily solvable.

Addressing the Arduino

If we want a simple way to know where data received by the XBee master device originated from, we can modify the Arduino code to contain unique identifying information. Let's look at code that has been modified to send back a unique ID along with its data packet:

```
int ledPin = 13;    // LED connected to digital pin 13
int read_val;    ❶

unsigned int IDval = 0x0011;   ❷

void setup()
{
    Serial.begin(9600);    // opens serial port, sets data rate to 9600 bps
        pinMode(ledPin, OUTPUT);      // sets the digital pin as output

}
```

```
    void loop() {
        if (Serial.available()) ❸
            {
            read_val = analogRead(A0) ❹
            Serial.print("ID = ");
            Serial.print(IDval); ❺
            Serial.print(" My Analog Is");
            Serial.println(read_val); ❻
        }
    }
```

❶   This variable stores values read from the Arduino analog pins.

❷   This is the unique ID that we've given to this particular Arduino. This ID should be different for each Arduino on the network.

❸   Ths line of code pauses checks for incoming data sent by the XBee master device. If the Arduino has received serial data, the `Serial.data()` command will return true, and we will execute the code in the `if` statement.

❹   We are reading and storing an analog value to send to the master device.

❺   This line of code causes the Arduino to send its unique ID back to the master device.

❻   In general, we can send any data we want back to the master device. This code transmits the most recently read analog value.

Whenever the XBee master device broadcasts a byte to all the sensor motes, the Arduino code running on the sensor motes will detect that byte and transmit a response. All of the sensor motes will send a response. The data contained in each response will reflect the analog value of the particular sensor mote that sent the data. Each sensor mote will also preface that data with a unique header, reflecting the ID value that we programmed into each Arduino.

The key to this technique is remembering to change the ID value each time you download the code into your sensor motes. This step is easy to forget if you are programming a large number of sensor motes.

One nice feature of this technique is that, by adding only three lines of code, we have added identifying data to the packets that are received by the XBee master device. We can also easily change the ID of each of the sensor motes by changing one line of the Arduino code.

Individual Call and Response

We just covered a basic technique for adding a unique address to the Arduino so that wireless packets can be distinguished from each other. When the XBee master device sends out any byte to the sensor motes, they all respond, however now they have a unique ID that we can use to figure out which device sent which data.

If we want an *individual* sensor mote to have an *individual* response, then we need to further modify our Arduino code. Fortunately, we already have the basic ingredients built into our last code example. Rather than just waiting for any byte to trigger a response, we can change our Arduino sensor mote code to detect our unique ID value. This requires detecting a multi-byte packet over the serial port:

```
int ledPin = 13;    // LED connected to digital pin 13
int read_val;

int SerID_In = 0;   ❶

unsigned int IDval = 0x0011;   ❷

void setup()
{
    Serial.begin(9600);    // opens serial port, sets data rate to 9600 bps
        pinMode(ledPin, OUTPUT);    // sets the digital pin as output

}
void loop() {
    if (Serial.available() > 1)   ❸
        {
           SerID_In = Serial.read();   ❹

           SerID_In = (SerID_In << 8) + Serial.read();   ❺

           If(SerID_In == IDval)   ❻
           {
               read_val = analogRead(A0);
               Serial.print("ID = ");
               Serial.print(IDval);
               Serial.print(" My Analog Is");
               Serial.println(read_val);
           }
      }
}
```

❶ This variable is used to store an ID value sent by the master to all the Arduinos on the network. This value represents the unique ID of the particular Arduino from which the master device is requesting data.

❷   The unique ID of this sensor mote is set here in the Arduino code. It should be different for each Arduino on this network.

❸   `Serial.available()` tells us how many bytes have been received. We will only take action and respond to serial data if we've received at least 2 bytes.

❹   Read the first byte from the serial port. This will become the most significant byte of the ID.

❺   Read the next byte from the serial port. This will be our least significant byte. Since `Serial.read()` only reads one byte at a time, we need to rebuild our data using left shifting and addition operations to combine the MSB and LSB into an integer value.

❻   Here, we compare the requested ID against our own unique ID. IF the values match, we will respond to the request by sending our data.

This code contains several improvements over our basic Arduino sensor mote ID code. First of all, we can uniquely send messages to each sensor mote individually. This is similar to the Unicast transmission mode, except that the messages from the XBee master device are being broadcast to all the sensor motes. Only the mote that sees its address will respond to the message.

Also, this code is superior to Unicast mode. We can change which sensor mote we are communicating with just by changing the contents of our outgoing serial packet. If we wanted to change sensor motes while using Unicast mode, we would have to reprogram the XBee master device and change the destination node address (DL) each time we communicated with a different sensor mote. Another serious benefit is that we have reduced the traffic on our wireless sensor network. Only the device from which we have requested data will respond to that request.

Now that we have a mechanism for sending different messages from the XBee master device to the Arduino sensor motes, we can easily add code that can product a different response to the various packets, as the sample below shows:

```
int ledPin = 13;    // LED connected to digital pin 13
int read_val;

int SerID_In = 0;
int Msg_In = 0;   ❶

unsigned int IDval = 0x0011;

void setup()
{
    Serial.begin(9600);       // opens serial port, sets data rate to 9600 bps
        pinMode(ledPin, OUTPUT);      // sets the digital pin as output
}
```

```
void loop()
{
    if (Serial.available() > 2)  ❷
    {
      SerID_In = Serial.read();

             SerID_In = (SerID_In << 8) + Serial.read();

      Msg_In = Serial.read();  ❸

      if(SerID_In == IDval)
      {
        switch (Msg_In)  ❹
        {
          case 0:
             Serial.println("Hello");  ❺
              break;
          case 1:
                Serial.println("World");  ❻
                break;
          default:
          {
             read_val = analogRead(A0);
             Serial.print("ID = ");
             Serial.print(IDval);
             Serial.print(" My Analog Is");
             Serial.println(read_val);
             break;
          }
        }
      }
    }
}
```

❶ We will use the variable `Msg_In` to store the message component of the data transmitted by the master.

❷ Now that we've added a message, we need to wait for 3 bytes. The first two bytes are ID bytes, and the third byte is the message that was sent.

❸ Here we grab the message byte. This byte stores a value that will determine what the mote's response will be.

❹ We receive a switch-case statement to execute a particular action, determined by the numerical value of the message received by the Arduino.

❺ A message of value "0" causes the Arduino to send the message `"Hello"` back to the master device.

❻ A message of value "1" causes the Arduino to send the message `"World"` back to the master device.

Once you've entered the code, plug your Arduino into your computer, then go ahead and compile and upload the sketch to your board.

 See the sections "Connecting to the Board" on page 5 and "Uploading the Sketch" on page 10 in Chapter 1 if you're having trouble uploading the sketch to your Arduino board.

Then, using a serial Terminal program like CoolTerm that allows you to send hexadecimal data to the serial port, send the *hexadecimal* values `00 11 00`, as shown in Figure 8-1, to test the code.

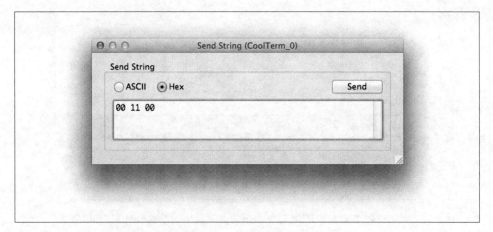

Figure 8-1. This is the Hex data to send via the XBee master

When your code is running successfully, you will see a `Hello` response, as shown in Figure 8-2.

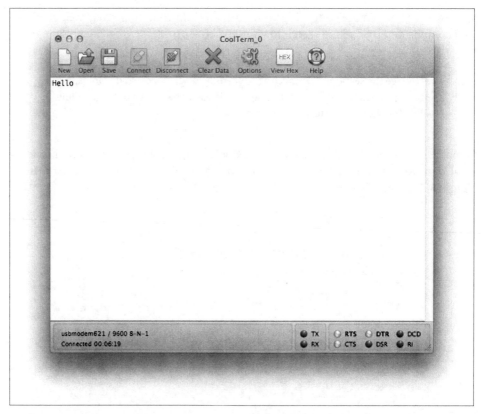

Figure 8-2. The sensor mote response generated from the XBee master command

If we wanted to request data from other devices, we can change the unique identifier from *0x0011* to *0x0012*. Also, if we wanted to change the data that we requested from a particular device, we could change the third byte in the packet from *0x00* to *0x01*, or whatever other values we've programmed into the Arduino sensor mote.

 The XBee master device can repeat the data request continuously until it gets a response or reaches a timeout value. This is one way, using application software rather than the Unicast firmware, to increase the robustness of your network communications.

Switching to Mesh Networks

We have shown here how to create a simple *star* or *master-slave topology* sensor network that uses a single master, connected to your computer. Despite its simplicity, there are a number of limitations to this technique.

We are using the processing power of the Arduino to detect messages containing an address that matches a pre-programmed unique identifier. This requires that we update the identifier in the Arduino IDE and recompile our code before individually programming each device. For a large network, all of this manual work would quickly become tedious.

Also, the star topology network can only grow in radius to be as big as the maximum point-to-point transmission distance of the XBee Series 1 devices. We have no ability to extend the radius of the network by hopping messages along a spread-out chain of XBee devices.

These problems can be addressed by switching to a *mesh network topology* using the XBee Series 2 devices. If that is your interest, check out *Building Wireless Sensor Networks* by Rob Faludi (O'Reilly).

Summary

In this chapter we incorporated multiple Arduino sensor motes into a many-to-point network, modified our sensor mote code to add a unique address to each Arduino, and then further modified the code to detect a multi-byte packet over the serial port, creating the capability for call and response with the master.

CHAPTER 9
Visualizing with Processing

Processing is a open source programming language and development environment built by—and for—the art and design community, to help teach the fundamentals of computer programming in a visual context, and to allow the community to more easily build interactive visualizations.

Processing

The Processing user interface is going to look hauntingly familiar (see Figure 9-1).

Figure 9-1. The Processing development environment

The Processing environment spawned several other projects, including Wiring and the Fritzing environment (which we met in Chapter 5), and was used as the basis for the Arduino development environment. There are still a lot of similarities between the two, and the familiarity you now have with the Arduino development environment should help you rapidly get up to speed with Processing.

Installing the Software

Download the latest version of the development environment from the Processing.org website (*http://processing.org/download/*).

 The latest version of the Processing Development Environment (PDE) is Processing 2.0b7, as of the time of this writing.

The development environment comes as zipped file, which should decompress automatically after you download it. If it doesn't, double-click on it to open it manually. After it is open, just drag the *Processing.app* application into your */Applications* folder.[1]

Reading Data From a File

You should already have some data from the sensor platform on disk in CSV format. It's fairly easy to read and parse that directly into the Processing environment. We're going to assume that you've currently got some data that looks a lot like the following:

```
22.60, 42.50, 14, 0, 0   ❶
21.50, 42.60, 25, 1, 1
23.60, 42.60, 30, 1, 0
24.70, 42.70, 114, 1, -1
25.60, 42.70, 20, 0, 0
22.60, 42.50, 14, 0, 0
21.50, 42.60, 25, 1, 1
13.60, 42.60, 30, 1, 0
14.70, 42.70, 114, 1, -1
15.60, 42.70, 20, 0, 0
     .
     .
     .
25.60, 42.70, 20, 0, 0
```

---

1. Full instructions for getting started with Processing on Windows, Linux, and Mac can be found here (*http://processing.org/learning/gettingstarted/*).

❶   From left to right, we're expecting temperature (in C), humidity (in %), the reading from the microphone, the current state of the PIR sensor, and markers denoting the start (1) and end (-1) of movement events.

Like Arduino sketches, Processing sketches consist of two main functions: the set up( ) function, and the draw( ) function, equivalent to the Arduino environment's loop( ). Go ahead and open Processing and enter the following code:

```
String[] lines;

void setup () {
  lines = loadStrings("file.dat");❶
  println("There are " + lines.length + " lines");
  for (int  i= 0; i < lines.length; i++) {
    float temp, humidity;
    int mic, pir, motion;
    if (lines[i] != null) {
      String [] numbers = split(lines[i],',');❷
      temp = float(numbers[0]);
      humidity = float(numbers[1]);
      mic = int(float(numbers[2]));
      pir = int(float(numbers[3]));
      motion = int(float(numbers[4]));

      print( "T = " );
      print( temp );
      print( "C, H = " );
      print( humidity );
      print( "%, S = " );
      print( mic );
      print( ", P = " );
      print( pir );
      print( ", M = " );
      println( motion);

    }
  }
}

void draw () {❸
  // do nothing
}
```

❶   We load the CSV file into an array of strings. Each line of the file will be stored as a separate string in the array.

❷   We loop around the array of strings and split each line on the position of the comma between the separated values. This is read into a separate array of strings, which we then convert to `floats` and `ints` appropriately and print them to the Notification Area of the Processing environment.

❸   Since the file already exists, we only want to read it once. The `draw( )` loop is therefore empty and we read the file during the `setup( )` function.

Save the Processing sketch, and drop your CSV file into the sketch's directory along with the *.pde* file so that it can find your data, then click on the Run button. You should see something like Figure 9-2: your data is being parsed and printed to the Notification Area in the development environment.

Figure 9-2. Reading data in from a file

From here, it's actually pretty easy to make a simple X-Y plot of value against time. To simplify things even further, we're going to start out by plotting only a single variable.

 The (0,0) coordinate of the Processing canvas is located in the upper-left corner. That means any (x,y) coordinate pairs are referenced to that point.

```
String[] lines;
float[] temp;
float xscale;

void setup () {
  size(820, 600); ❶
  background(0);

  PFont f = createFont("Arial",16,true; ❷
  textFont(f,16);
  fill(255);

  stroke(255); ❸
  line( 25, 25, 25, 575 );
  line( 25, 575, 775, 575 );
  line( 775, 575, 775, 25 );
  line( 775, 25, 25, 25 );

  lines = loadStrings("file.dat");
  println("There are " + lines.length + " lines");
  xscale = 750.0/lines.length;
  println("The xscale is therefore = " + xscale );

  temp = new float[lines.length];
  for (int  i= 0; i < lines.length; i++) { ❹
    if (lines[i] != null) {
       String [] numbers = split(lines[i],',');
       temp[i] = float(numbers[0]);
    }
  }

  for (int i = 0; i < lines.length; i++) {
    stroke(204, 102, 0); ❺
    strokeWeight(1.2); ❻
    float ty = map( temp[i], 15.0, 50.0, 25.0, 575.0 ); ❼
    ty = 575.0 - ty; ❽
    float tx = 25.0 + (float)i * xscale; ❾
    point(tx,ty); ❿

  }

  text("15", 5, 580); ⓫
```

Reading Data From a File | 123

```
    text("C", 10, 300);
    text("50", 5, 35);
  }

  void draw () {
    // do nothing
  }
```

❶ Here we create a canvas that is 820 pixels wide by 600 pixels high, and in the following line set the background color to black.

❷ We're going to draw some labels later on, so we need to create a font to do that. Here we set the default font to be 16 pt Arial.

❸ Here we set the line color to white, then draw a rectangular box on the canvas to frame our graph and act as axes. The canvas is 820×600 pixels; our axes are going to have a top-left corner at (25, 25) and a bottom-right corner at (775, 575). The canvas area we're going to use for our plot is therefore 775 - 25 = 750 pixels wide, and 575 − 25 = 550 pixels high.

❹ This is the same loop as we had in the previous code sample, however here we're just pulling the first number from each line: the temperature.

❺ Looping through each point, we set the color of the point to be a pleasant orange so that it's distinct from the white axes.

❻ We make the point about 20% larger than a single pixel.

❼ We use Processing's map( ) function to map the numbers in the temp[ ] array, which should be in the range from 15–50 (as they are in centigrade) to numbers more appropriate for the size of our canvas. We're going to map 15 to 25 and 50 to 575 and scale the numbers in the array appropriately.

❽ Since the (0,0) coordinate of the Processing canvas is in the upper-left corner, we need to flip our plot upside down, so that our numbers appear the right way up.

❾ Here we could have again used a map( ) function, but it was easier to manually scale each point to the size of our plotting area manually.

❿ Using our newly scaled x and y positions, which are now in real pixels relative to the upper-left corner of the plot, we plot a point and loop back around to start again.

⓫ Here we put some really simple labels on our y-axis to show the range in temperatures we're plotting to the canvas.

Save the Processing sketch—making sure your CSV file is saved into the sketch's directory along with the *.pde* file—and click on the Run button again. You should see something like Figure 9-3.

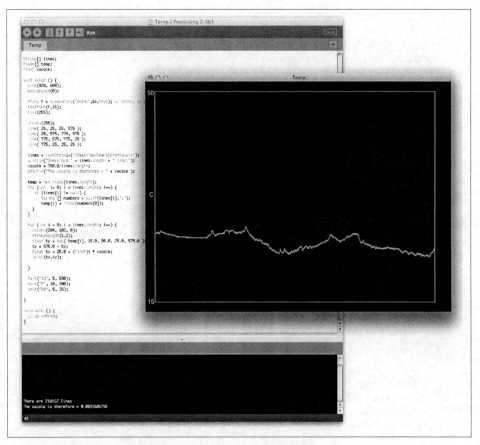

Figure 9-3. Graph of temperature data against time

 The temperature data shown in Figure 9-3 is actual data taken in New York during Hurricane Sandy at the tail end of 2012 (*http://goo.gl/ KbA0J*).

From here, it's simple to move forward and plot some of the other columns on top of the same graph. Go ahead and make the following changes, shown in bold, to your code. We're going to read and plot the humidity as well as the temperature on our graph:

```
String[] lines;
float[] temp;
float[] humidity;
float xscale;

void setup () {
  size(820, 600);
```

Reading Data From a File | 125

```
  background(0);

  PFont f = createFont("Arial",16,true); // Arial, 16 point, anti-aliasing on
  textFont(f,16);
  fill(255);

  stroke(255);
  line( 25, 25, 25, 575 );
  line( 25, 575, 775, 575 );
  line( 775, 575, 775, 25 );
  line( 775, 25, 25, 25 );

  lines = loadStrings("file.dat");
  println("There are " + lines.length + " lines");
  xscale = 750.0/lines.length;
  println("The xscale is therefore = " + xscale );

  temp = new float[lines.length];
  humidity = new float[lines.length];
  for (int  i= 0; i < lines.length; i++) {
     if (lines[i] != null) {
        String [] numbers = split(lines[i],',');
        temp[i] = float(numbers[0]);
        humidity[i] = float(numbers[1]);
     }
  }

  for (int i = 0; i < lines.length; i++) {
    stroke(204, 102, 0);
    strokeWeight(1.2);
    float ty = map( temp[i], 15.0, 50.0, 25.0, 575.0 );
    ty = 575.0 - ty;
    float tx = 25.0 + (float)i * xscale;
    point(tx,ty);

    stroke(204, 0, 0);
    float hy = map( humidity[i], 0.0, 100.0, 25.0, 575.0 );
    hy = 575.0 - hy;
    float hx = 25.0 + (float)i * xscale;
    point(hx,hy);

  }

  text("15", 5, 580);
  text("C", 10, 300);
  text("50", 5, 35);

  text("0", 780, 580);
  text("%", 780, 300);
  text("100", 775, 35);

}
```

```
void draw () {
  // do nothing
}
```

Make sure you save the Processing sketch, and click on the Run button again. You should see something like Figure 9-4. As you can see, we used the righthand y-axis for the scale for our humidity measurements.

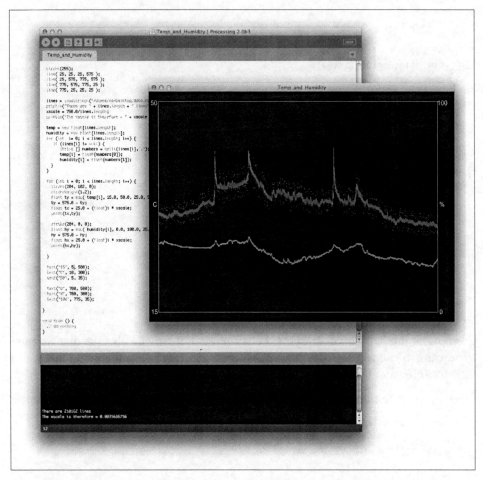

Figure 9-4. Graph of temperature and humidity data against time

Reading Data Directly From the Serial Port

Reading data from an existing file is all very well, but we really want to display our data in real time, as it arrives from the sensor platform. Fortunately, that's pretty easily done with Processing.

Go ahead and start a new Processing sketch and add the following code:

```
import processing.serial.*;

Serial port;

void setup () {
   println(Serial.list());
}

void draw () {

}
```

Save your sketch and click on the Run button. You should get a list that looks a lot like the one below, appearing in the Notification Area:

```
Stable Library
=========================================
Native lib Version = RXTX-2.1-7
Java lib Version   = RXTX-2.1-7
[0] "/dev/tty.Bluetooth-PDA-Sync"
[1] "/dev/cu.Bluetooth-PDA-Sync"
[2] "/dev/tty.Bluetooth-Modem"
[3] "/dev/cu.Bluetooth-Modem"
[4] "/dev/tty.usbmodem401211"
[5] "/dev/cu.usbmodem401211"
```

From using the Arduino development environment, it may be obvious that we want Processing to connect to /dev/tty.usbmodem401211 (port 4) in the list so that it can talk to our sensor platform.

Open a new sketch and add the following code—here, we're replicating our first sketch:

```
import processing.serial.*;

Serial port;          // The serial port

void setup () {
  size(800, 600);
  port = new Serial(this, Serial.list()[4], 9600); ❶
  port.bufferUntil('\n'); ❷
  background(0);
}

void draw () {
```

```
  }
  void serialEvent (Serial port) {❸
    String inString = port.readStringUntil('\n');

    float temp, humidity;
    int mic, pir, motion;
    if (inString != null) {
      String [] numbers = split(inString,',');
      temp = float(numbers[0]);
      humidity = float(numbers[1]);
      mic = int(float(numbers[2]));
      pir = int(float(numbers[3]));
      motion = int(float(numbers[4]));

      print( "T = " );
      print( temp );
      print( "C, H = " );
      print( humidity );
      print( "%, S = " );
      print( mic );
      print( ", P = " );
      print( pir );
      print( ", M = " );
      println( motion);

    }
  }
```

❶   We open a new serial connection to the serial port at 9600 baud.

❷   We'll buffer any characters coming from the serial port until we see a carriage return (\n) character, at which point we'll call the serialEvent( ) function.

❸   When we receive a \n character, the serialEvent( ) function will be called by the Processing sketch; this will start the parsing of the CSV in the "line" we have received over the serial connection. We do this in exactly the same we did to the lines in our CSV file at the start of the chapter.

If all goes well, you should see that after a while the parsed sensor readings start to appear in the Notification Area.

Plotting Temperature in Real Time

From here, it's not too hard to write a sketch to plot the incoming data in real time. For simplicity's sake, we're going to look at plotting just temperature.

We're going to establish an initially "empty" array of values, and then make use of the draw( ) function to continuously plot this array to our canvas.

At the same time, we'll be monitoring the serial port. When we receive data, the `serialEvent()` function will be called by the sketch, the contents of the array will be shuffled along, and the incoming data will be added to the end of the array. Over time, old data will drop off the "front" of the array, and new data will continue to arrive at the "end," and you'll end up with a graph that is constantly updated by the incoming data.

```
import processing.serial.*;

Serial port;
float xscale;
float[] values = new float[64]; ❶

void setup() {
  size(820, 600);
  println(Serial.list());
  port = new Serial(this, Serial.list()[4], 9600);
  port.bufferUntil('\n');

  xscale = 750.0/64; ❷

  PFont f = createFont("Arial",16,true); // Arial, 16 point, anti-aliasing on
  textFont(f,16);
  fill(255);

}

void draw() {
  background(0); ❸
  stroke(255);
  strokeWeight(1);
  line( 25, 25, 25, 575 );
  line( 25, 575, 775, 575 );
  line( 775, 575, 775, 25 );
  line( 775, 25, 25, 25 );

  text("20", 5, 580);
  text("C", 10, 300);
  text("40", 5, 35);

  for (int i = 0; i < 63; i++) { ❹
    stroke(204, 102, 0);
    strokeWeight(5);
    float ty = map( values[i], 20, 40, 25.0, 575.0 );
    ty = 575.0 - ty;
    float tx = 25.0 + (float)i * xscale;
    point(tx,ty);
  }

}

void serialEvent(Serial port) { ❺
  String inString = port.readStringUntil('\n');
```

```
        float temp;
        if (inString != null) {
          String [] numbers = split(inString,',');
          temp = float(numbers[0]);
          for (int i = 0; i < 63; i++) {❻
             values[i] = values[i + 1];
          }
          values[63] = temp;❼
          println( "T = " + temp + "C" );
        }
    }
```

❶ We assign an empty array of size 64 to hold the incoming data. Initially, all the values in this array will be zero.

❷ As before, we have 750 pixels inside our axes to plot our data. However, this time we know that we'll have 64 data items to plot, so our x-scale is therefore going to be 750/64 = 1.171875.

❸ Calling `background(0)` at the start of the `draw()` function clears the canvas and removes all the old data points. This does mean that we have to redraw our axes each time through the function, and we do that over the course of the following few lines.

❹ This `for()` loop goes through and plots each of the 64 data points in the array to the canvas. Initially, this array will just contain zeros.

❺ The `serialEvent()` function will be called when we get a "line" of data from the serial port.

❻ This `for()` loop shuffles the data currently in our array one slot lower, dropping `values[0]` off the bottom of the array, and leaving the final item in the array as a duplicate that we can then overwrite with new data.

❼ Here we overwrite the new incoming temperature to the last item in the array; next time through, the `draw()` routine will use this value to build our graph.

Save the Processing sketch and click on the Run button again. You should see something like Figure 9-5, with new data arriving on the righthand side of our plot and the graph moving leftwards as time progresses.

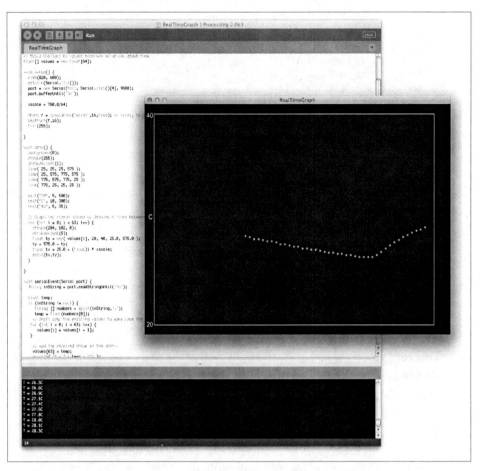

Figure 9-5. Reading real time data from the serial port

Summary

We've just scratched the surface. Processing is capable of much more than simple X-Y plots, including building complex 3D interactive visualizations. If you're interested in learning more about Processing, see Chapter 11 for a list of additional material.

CHAPTER 10
Visualizing with LabVIEW

LabVIEW is a graphical programming tool and development environment that was developed by the National Instruments Corporation. It is commonplace in settings that rely on automation and control hardware and software. LabVIEW (which stands for *Laboratory Virtual Instrumentation Engineering Workbench*) is also used extensively in academic settings, especially in research labs that rely on custom instrumentation. FIRST Robotics competitors will recognize LabVIEW as the tool that is used to program their robots as the alternative to C or C++. Also, LabVIEW is the software that drives the LEGO Mindstorm NXT product.

Although LabVIEW is not open source, we highlight it here as an interesting alternative approach to creating a graphical, instrument-like user interface for your projects.

LabVIEW

The LabVIEW development environment is fundamentally different than the Processing-derived environments like Fritzing or Arduino that we covered in earlier chapters. A basic LabVIEW program consists of a graphical user interface (GUI) known as the *Front Panel* and a diagram where the software is built using graphical code blocks, known as the *Block Diagram* (see Figure 10-1).

The LabVIEW environment might seem like a radical departure from the familiar text-based programming languages and environments that we've been using up to this point. Graphical programming does take a little time to get comfortable with. As we go through this chapter, keep in mind that any new programming language takes time to get used to. With LabVIEW, we have to get used to seeing graphical representations of our program elements, versus seeing our program elements represented as text.

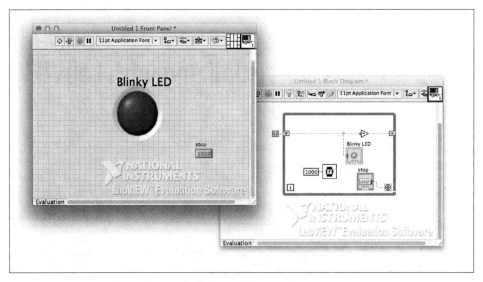

Figure 10-1. Screenshot of a simple LabVIEW program

Installing the Software

Download an evaluation version of the LabVIEW environment from National Instrument's website (*http://www.ni.com/trylabview/*).

 You will need to create an account at National Instruments in order to download the evaluation. At the time of this writing, the latest evaluation version of LabVIEW is 2012.

The download file is rather large—slightly over 1 GB for the PC version and over 600 MB for the Max OS X version. When it the download is complete, double-click on the install file. On the Mac, this will open *LabVIEW2012Evaluation.dmg*. Once you've opened up the disk image, you will want to install *LabVIEW 2012 Evaluation.mpkg*.

If you are running Mac OS X before 10.8, simply double-click this file to install it. If you are running Mac OS X 10.8 or later, you will need to bypass *Gatekeeper*, the built-in software designed to keep malicious software off your computer. The evaluation software does not have the official Apple software developer signature, so it won't be automatically recognized as a safe file. To bypass *Gatekeeper*, control-click on the *LabVIEW 2012 Evaluation.mpkg* file. It will open to the installer (Figure 10-2). Click *Continue* to begin installation.

Figure 10-2. The LabVIEW installation window

For PC downloads, the installation should begin automatically. If it doesn't, navigate to the downloaded file and double-click on it to launch the installer manually. Once you've completed the installation, you will need to restart your computer.

After you restart your computer, navigate back to the *Drivers* folder in the install file, and install the *VISA* drivers. You will also need to restart your computer after these drivers are installed. We will need these drivers to communicate between LabVIEW and our Arduino.

Once you've properly installed all the LabVIEW files, you can double-click the application to run it. You will get a window that looks like Figure 10-3.

Figure 10-3. The LabVIEW startup window

Let's go ahead and create our first project by clicking on the *Create Project* button. You will be presented with a list of starting points for your project. We will select the Blank VI template to get started (see Figure 10-4).

Click Finish to see your new *Blank VI* windows pop up.

 We call this program a *VI* beacuse LabVIEW was originally developed to make it easy for scientists to quickly create *Virtual Instruments*, software with GUIs that copied the look and feel of a physical instrument being automated or replaced by the LabVIEW program.

Figure 10-4. Selecting the Blank VI template

Our Blank VI starts out as an empty user interface *Front Panel* and blank *Block Diagram* that we can add our code into (see Figure 10-5).

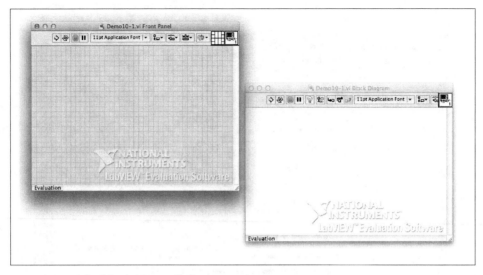

Figure 10-5. The blank VI, ready for programming

Our first program will be a fairly standard exercise. We will make a virtual LED blink once per second. We need to add an LED to our front panel. Right click anywhere on the front panel to bring up the Controls Palette (see Figure 10-6).

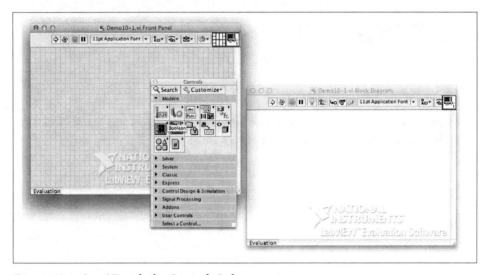

Figure 10-6. Our VI with the Controls Palette open

Move your mouse over the top row of icons. If you mouse over the second column, you will see the word *Boolean* pop up. This is where front panel objects that can control (input) or display (output) Boolean values can be found. If you click on this icon, the controls palette will change to display only Boolean input or output items.

Click on the simple round LED. Your mouse will change to a hand, and you can place the LED anywhere on your front panel. Go ahead and place a single LED on your front panel. Also, place one of the **Stop** buttons on your front panel.

We now have an input and output for our program. You will also notice that two green-bordered squares have just appeared on our block diagram (see Figure 10-7). These are the variables that control or read out the state of the objects that we've just placed on our front panel.

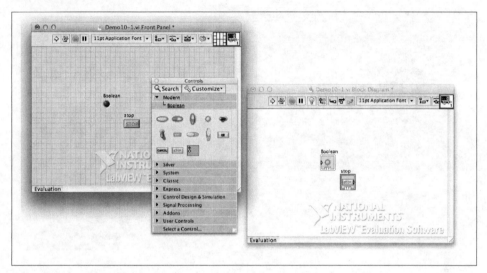

Figure 10-7. We've just placed our first input and output

 We can tell which variables on our block diagram are inputs by the small arrow to the right of the icon, meaning data input to the front panel will "flow" into the program from the front panel object. These inputs also have a thicker boarder around the icon. Icons that are outputs to the front panel have a thinner outline, and there is a small arrow indicating that data will flow from the block diagram out to the front panel object.

If you want to change the size of your LED, you can drag the corners of the icons for the LED and stop button that you've placed to change their sizes. Go ahead and try it. Drag the corners of your LED to make a large, dramatic LED. Also, double-click on the name *Boolean* to change it to **LED** so our LED is labeled appropriately.

Once you are satisfied with the size of your LED, let's look more closely at the block diagram. You can see that our program consists of two variables, and nothing more. These icons represent the code that connects the state of the front panel object to the value of the variable. If we write the value of **True** to the LED, then the LED should turn on. Let's try this.

Right-clicking in the block diagram will bring up the Functions palette. Click on the upper-rightmost icon in the palette. These are our Boolean-related functions (see Figure 10-8).

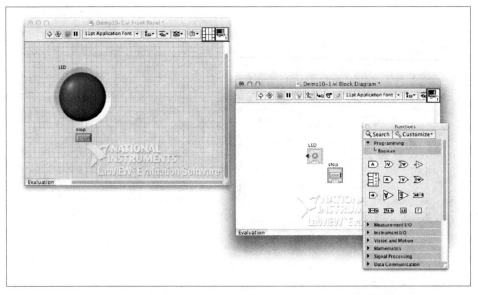

Figure 10-8. We have a big LED, and our Functions pallet is on Boolean

Click on the icon marked with the letter T for True and drop the icon on your block diagram to the left of the LED. This is a program constant that we will associate with our LED. Hover your mouse over the right side of the True constant. Your mouse will momentarily transform to look like a spool of wire. Click on the right side of the True constant and drag a wire to the center-left of the LED. You should see the wire automatically jump onto the center of the LED icon. Release the mouse button.

Congratulations! You have just created your first LabVIEW program.

Granted, this is a very simple program. When we run it, the only action that our program will perform is to pass the value of the True constant over to the LED.

Let's make sure this simple program works. Locate the white arrow in the upper-left corner of either the front panel or block diagram. This is the Run button (see Figure 10-9). Click this button to run the program.

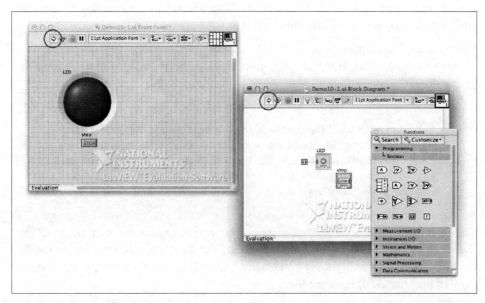

Figure 10-9. The Run button

When you click the *Run* button, your LED should change from dark green to light green, indicating that it has been turned on.

Congratulations again! You've just run your very first LabVIEW program.

Now, this isn't a very interesting program. We'd like to make our LED blink. To do this, we need to add a few more program elements. If we were using a standard text-based language like Arduino, a program that blinks an LED might look like the following:

```
// the loop routine runs over and over again forever:
void loop() {
    digitalWrite(led, HIGH);    // turn the LED on (HIGH is the voltage level)
    delay(1000);                // wait for a second
    digitalWrite(led, LOW);     // turn the LED off by making the voltage LOW
    delay(1000);                // wait for a second
}
```

To blink an LED in LabVIEW, we will need to find a loop command and a delay command, and we will need to set the LED high or low.

We find the loop command the same way we found the True constant. Navigate through the menus in the Functions palette until you find the *Programming>Structures* subpalette, and select the icon labeled While loop. Your mouse will turn into an icon that looks like a tiny loop. Now click near the upper-left corner of your block diagram and drag towards the lower-right corner, making sure that the dotted lines of the box that you are now drawing encircle the LED and Stop icons. Click the mouse to complete your box. You've now placed a while loop into your program.

Return to the Functions palette and locate the *Programming>Timing* sub-palette. Find the icon that says Wait, and place it in your block diagram, inside the while loop that you've drawn.

 As you begin to feel comfortable navigating around the various Functions palettes, you'll notice how similar this feels to building our Fritzing breadboard diagram. Also, the Wait icon that we've placed is a function that behaves exactly like the Arduino `delay()` function, so that should be familiar territory as well.

The Wait function, like `delay()`, causes our program to pause for the number of milliseconds that is passed to the function. To set the length of our wait, locate the *Programming>Numeric* sub-palette and place a Numeric constant on your block diagram close to the left side of the Wait icon.

If you place the constant close enough to the icon, you'll notice that a small wire is automatically created, connecting the constant and the Wait icon. If that didn't happen automatically, that's okay. Just click near the right side of the constant; your mouse will turn into the wire spool. Manually draw a wire from the constant to the Wait function. Now your LabVIEW program should look something like Figure 10-10.

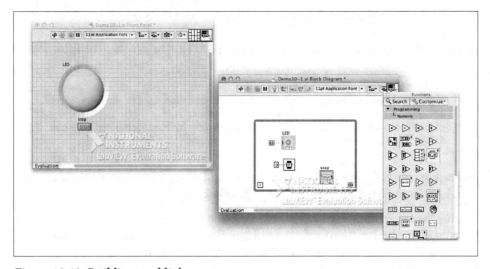

Figure 10-10. Building our blink program

You will notice two things. First, the run button looks like a broken arrow. That indicates that there is an error in our program, and it won't run successfully. Second, there is a small orange dot on the left side of the Wait function. That dot indicates that we are passing the wrong data type into the Wait function.

If you right-click on the constant, you will get a menu called *Representation* that shows a list of options for the various data types available for this constant. Set the constant to be a 32-bit unsigned value by selecting *U32* from the menu. The dot should go away.

The run arrow is still broken. LabVIEW is a strongly typed language, meaning that incompatible types will essentially cause an error, as represented by the broken arrow. In this case, however, the default type of the constant, a 32-bit signed value I32, is compatible with the U32 input of the Wait function. LabVIEW has automatically cast the I32 to a U32 for us, and indicated the cast with the orange dot that we eliminate by correcting the type of the constant. The broken arrow is representing another problem.

In the lower-right corner of the While loop, there is a small red stop sign. Remember that a While loop needs a stop condition. LabVIEW detects this, and won't let us run our program until we send a stop value to the While loop.

Connect a wire from the Stop icon to the stop sign in the corner of the While loop. Now our arrow will turn solid white, indicating that our program has no errors and is ready to run.

A quick trick for finding errors in your program is to click on the broken arrow. LabVIEW will bring up a detailed list of errors preventing the program from running. If you click on a particular error in the list and click the Show Error button, LabVIEW will highlight the problem in your block diagram, clearly showing what you need to fix.

Before the program will blink our LED, we need to make a few other changes. Double-click on the numeric constant wired to the Wait function. The number in the middle of the constant will be highlighted. Change this number from **0** to **1000**. We want a one-second—or 1000-millisecond—blink rate, so we need to set the Wait function to delay the program for 1000 milliseconds each time through the While loop.

Now, drag the True constant to the outside of the While loop. We want the LED to blink, not stay on continuously, so we don't want to set the LED to True every time through the loop. We do want to set the initial value of the LED to be True, and then each time through the loop we will invert the value of the LED. In the Boolean palette where you found the True constant, locate the Not function and drag it onto the block diagram inside the While loop. Wire the output of the Not function to the input of the LED.

We are almost done. We just need a variable that can store the last value of the Not function each time through the loop so that we can invert the LED value compared to its previous loop value, causing it to blink. LabVIEW has a handy capability built into every loop. If you right-click anywhere on the left or right side of the loop, you will see a menu with an item that says *Add Shift Register*. Go ahead and select this menu item. You will see the icon of a small black box with a black arrow that appears on the lefthand side of our While loop, and a corresponding icon on the righthand side of the While loop. This is a **Shift Register**. It acts as a local variable for the While loop.

Draw a wire from the True constant to the left shift register, and from the left shift register to the input of the Not function. Now draw a wire from the output of the Not function to the right shift register. If you've done this correctly, your program is ready to run and should look like Figure 10-11.

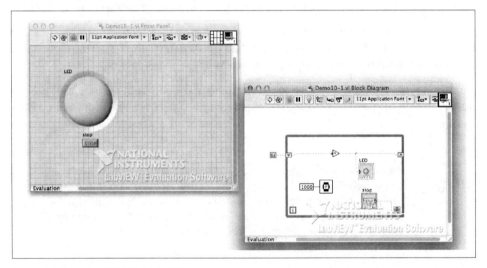

Figure 10-11. LabVIEW LED blink program

Click on the Run button to see your program run. You should see your LED blink. Click on the Stop button and change the value of your Wait function to **500**. The LED now blinks twice as fast, as we'd expect.

With the program still running, click the light bulb to the right of the Run arrow. When you do this, you'll see an animation of little dots traveling along your wires (Figure 10-12). You'll also see values next to our function's inputs and outputs. These dots represent the data moving from a variable to a function, or vice versa. The values you see are the actual data values that have been passed into or out of each element in our program.

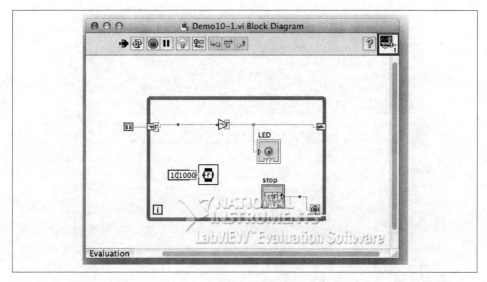

Figure 10-12. Our program in graphical debug mode

You have now built, run, and debugged your first full LabVIEW program. Our next step is to integrate Arduino communication into the program.

Simple LabVIEW with Arduino

We want to establish basic serial communications with our Arduino through LabVIEW. The best place to start is with a stock LabVIEW example. Navigate to the LabVIEW *Help* menu, and select *Find Examples*. A window will pop up with a list of LabVIEW example programs.

Click on the *Search* tab in the NI Example Finder window, and type `Serial` in the search box. The keyword *Serial* will appear in the search menu. Double-click on this keyword to show all the serial examples. In the example list that appears, double-click on *Basic Serial Write and Read.vi* (see Figure 10-13). This example will be the starting point for our Arduino communication program in LabVIEW.

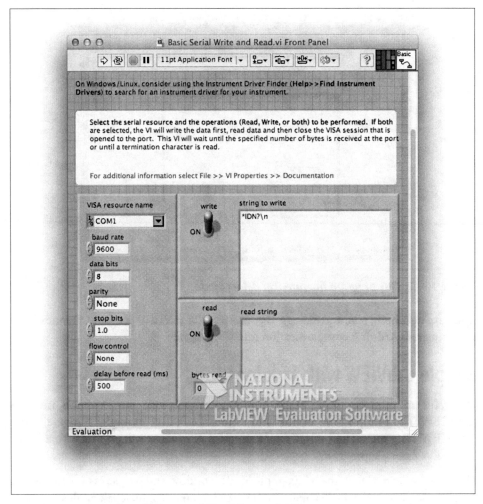

Figure 10-13. Basic Serial Write and Read Example VI

Now, take your Arduino and load some very simple code on it:

```
int ledPin = 13;     // LED connected to digital pin 13
int read_val;        // just use this for temporary storage of an analog value

void setup() {
    Serial.begin(9600);      // opens serial port, sets data rate to 9600 bps
    pinMode(ledPin, OUTPUT);       // sets the digital pin as output
}

void loop() {
    Serial.println("Hello From Arduino");
    delay(1000);
}
```

With that code loaded, you are almost ready to talk to your Arduino using LabVIEW. Like in the Arduino IDE, we need to select the serial port we want to use for our LabVIEW program. You'll notice a drop-down box labeled *VISA resource name* on our *Basic Serial Write and Read.vi* front panel. Click this drop-down box and select the serial port that most closely matches the port you had selected when you programmed your Arduino. Then click the Run button on the LabVIEW program. With any luck, you will read data from your Arduino into LabVIEW, and your LabVIEW front panel will look like Figure 10-14.

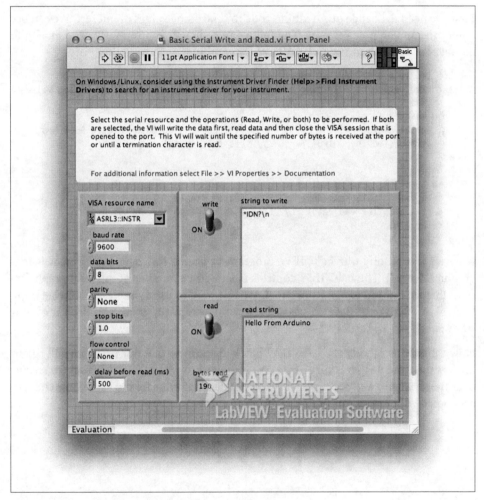

Figure 10-14. Reading from Arduino into LabVIEW

If your read string has extra characters in it, like \s or \r, right-click on the string display area and select *Normal Display* from the drop-down menu.

Graphing the Data

Now that we have basic communication established between our Arduino and our LabVIEW program, it would be nice to do something more interesting than sending "Hello From Arduino" continuously. With LabVIEW, it is very easy for us to graph data generated by the Arduino.

First, we need to change our Arduino code to send some values of interest. We can use one of the analog pins for now. The following code will continuously send analog to digital conversion readings to LabVIEW for graphing:

```
int ledPin = 13;    // LED connected to digital pin 13
int read_val;       // just use this for temporary storage of an analog value

void setup() {
    Serial.begin(9600);      // opens serial port, sets data rate to 9600 bps
    pinMode(ledPin, OUTPUT);    // sets the digital pin as output
}

void loop() {
  read_val = analogRead(A0);    // Read analog Pin A0
  Serial.println(read_val);     // Send that value up the serial port
  delay(100);
}
```

Now we can modify our LabVIEW program to accept this data and present it in a graphical format. The LabVIEW controls palette has a number of different graph formats that we can add to our front panel. We just want to use a simple Waveform Graph control to display our data, so we will select the waveform graph and drop it on our front panel.

Now we need to modify the LabVIEW block diagram so that our program will run as long as the Read switch is turned on. We also need to parse the string that we receive over the serial port and convert that string to a decimal value for display. Those changes are shown with comments in Figure 10-15.

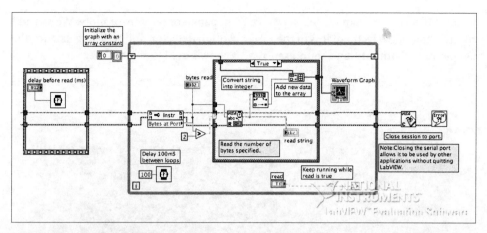

Figure 10-15. These modifications give us Arduino graphing

After making these changes to the Arduino code and the LabVIEW code, we can generate a graph of our Arduino's analog data (see Figure 10-16).

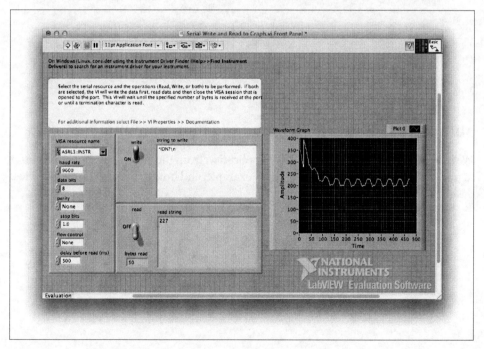

Figure 10-16. The Arduino A0 analog data is graphed

With a little work, we can make a very nice front panel for our sensor mote. We can add controls that show us the status of the LEDs, motion detector, and trending data for the temperature, humidity, and sound sensors, as shown in Figure 10-17.

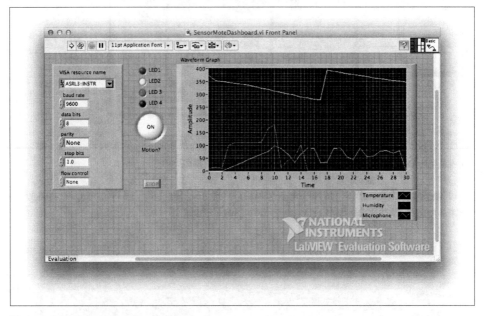

Figure 10-17. Sensor mote dashboard

Summary

We have seen how to build a LabVIEW application from the ground up, how to connect to our Arduino starting from a LabVIEW example, and how to make a simple graphical dashboard for our sensor mote project.

CHAPTER 11
Going Further

This book has walked you through getting started building your own hardware, taking data, and then doing some basic visualization on that data. But there is a lot more to learn.

Arduino

The Arduino is an amazingly flexible platform, and like several other topics in this book we really haven't gone into huge amounts of detail as to its capabilities.

If you want to learn more, you should probably take a look at the *Arduino Cookbook* (2nd ed.) by Michael Margolis (O'Reilly), or the *Arduino Up and Running* video course (*http://shop.oreilly.com/product/9780596803827.do*) with Brian Jepson (O'Reilly).

XBee Networking

If you're interested in building larger XBee-based networks, you might want to look at *Building Wireless Sensor Networks* by Robert Faludi (O'Reilly). While that book uses Series 2 modules exclusively, it can also help you with Series 1 networks, as Series 1 commands are—for the most part—just a subset of the Series 2 command set.

If you are interested in other projects that make use of the Series 1 modules we talked about in this book, check out *Making Things Talk* (*http://oreilly.com/catalog/0636920010920*) by Tom Igoe (O'Reilly).

Fritzing

We introduced you to Fritzing in Chapter 5, where we talked about laying out circuits in the Fritzing Breadboard view. However, we didn't really talk about circuit schematics, or how to layout custom PCBs. If you want to learn more about Fritzing, there are some really great online resources on their site (*http://fritzing.org/learning/*).

EAGLE

If you grow beyond what Fritzing is capable of, or if you just don't get on with it and you're looking for alternatives, you should probably take a look at EAGLE. It's a freely available and fairly powerful PCB design package that runs on Windows, OS X, and Linux.

EAGLE can be downloaded here (*http://www.cadsoftusa.com/download-eagle/*).

Processing

Processing is capable of much more than we've used it for so far. There have been some really stunning visualizations done using it. Check out *Getting Started with Processing* by Casey Reas and Ben Fry (O'Reilly Media).

ProcessingJS

In Chapter 9, we looked at the native version. However, there is also a port of Processing for JavaScript (*http://processingjs.org*). Processing.js is written in JavaScript, and uses HTML5's `<canvas>` element. It converts your Processing code to JavaScript and runs it in the browser. That means that you can experiment with your visualizations on the desktop and them drop them directly onto the web.

Quick Start

You can download the latest version of Processing.js here (*http://processingjs.org/download/*), and install the JavaScript library to your web site. Then go ahead and write your *.pde* sketch as you normally would, and test it locally in the normal Processing environment. Then copy your *.pde* file to your website and create a web page that includes the sketch and a `<canvas>` element to run it in:

```
<script src="processing-1.4.1.min.js"></script>
<canvas data-processing-sources="hello.pde"></canvas>
```

You can specify multiple *.pde* files in the canvas element by separating the file names with spaces. Then load your web page in a browser, and it will parse and run your sketch.

LabVIEW

If you are intrigued by the capabilities of LabVIEW, it is worth downloading the free LabVIEW Interface for Arduino (LIFA) from National Instruments (*http://sine.ni.com/nips/cds/view/p/lang/en/nid/209835*).

This toolkit boasts:

- Easy access to Arduino DIO, AI, PWM, I2C, and SPI from LabVIEW
- I/O engine sketch to load on Arduino
- Examples for basic tasks and sensors
- Wireless with Bluetooth or XBee
- Loop rates: USB tethered (200 Hz) and wireless (25 Hz)
- Open Arduino sketch & toolkit VIs help you customize functionality

Also, consider checking out a video walkthrough (*http://zone.ni.com/wv/app/doc/p/id/wv-1344/upvisited/y*) of the features and functions of LabVIEW.

Programming with graphical tools is definitely an acquired taste. But for many engineers, LabVIEW is the tool of choice for rapid prototyping and rapid development of instrumentation or control software. If you find yourself liking LabVIEW, you are not alone.

Data Visualization

We introduced you to creating basic graphs with Processing and LabVIEW, but there is so much more to explore when it comes to presenting your data to others (or yourself). Another popular tool is a JavaScript library called D3 (*http://d3js.org/*); if that seems up your alley, see *Getting Started with D3* by Mike Dewar (O'Reilly).

Data visualization is a deep and rich discipline of its own, but here are some resources you may want to have on hand, no matter what languages or platforms you're using.

For a behind-the-scenes collection of case studies from practitioners working on all kinds of visualization projects, check out *Beautiful Visualization*, edited by Julie Steele and Noah Iliinsky (O'Reilly Media).

Graph Design for the Eye and Mind by Stephen M. Kosslyn (Oxford University Press) covers best practices for many standard graphs and explains the cognitive principles at work in a concise volume. For an even more solid understanding of visual perception and cognition, it doesn't get any better than *Information Visualization: Perception for Design* by Colin Ware (Morgan Kaufmann).

The Practical Guide to Information Design by Ronnie Lipton (Wiley) is a straightforward book on best practices for many kinds of design elements—useful in data visualization and other forms of graphic design.

And of course, for a thorough, in-depth treatment of the subject by the godfather of data visualization, see Edward Tufte's books: *The Visual Display of Quantitative Infor-*

mation, Envisioning Information, Visual Explanations: Images and Quantities, Evidence and Narrative, and *Beautiful Evidence* (Graphics Press).

About the Authors

Alasdair Allan is the author of several books published by O'Reilly Media, including this one. He and Pete Warden (*http://petewarden.typepad.com/*) are somewhat infamous for causing a privacy scandal by uncovering that your iPhone was recording your location (*http://radar.oreilly.com/2011/04/apple-location-tracking.html*) all the time. This caused several class action lawsuits and a U.S. Senate hearing. He isn't sure what to think about that. From time to time he stands in front of cameras (*http://shop.oreilly.com/product/0636920026457.do*), and you can often find him at conferences (*http://lanyrd.com/profile/aallan/*).

He runs a small technology consulting business (*http://babilim.co.uk/*) writing software, building hardware, and providing training; including a series of workshops on sensors (*http://sensorworkshops.com/*). He sporadically writes blog posts (*http://dailyack.com/*) about things that interest him, or more frequently provides commentary about them in 140 characters or less (*http://twitter.com/aallan*).

Alasdair is also a senior research fellow (*http://emps.exeter.ac.uk/physics-astronomy/staff/aa247*) at the University of Exeter (*http://www.exeter.ac.uk/*). As part of his work there he built a distributed peer-to-peer network of telescopes (*http://www.estar.org.uk/*) which, acting autonomously, reactively scheduled observations of time-critical events. Notable successes included contributing to the detection of the most distant object yet discovered, a gamma-ray burster at a redshift of 8.2 (*http://arxiv.org/abs/0906.1577*).

Kipp Bradford is an educator, technology consultant, and entrepreneur with a passion for making things. He was the founder or cofounder of start-ups in the fields of transportation, consumer products, HVAC, and medical devices, and holds numerous patents for his inventions. Some of his more interesting projects projects have turned into Kippkitts (*http://kippkitts.com*).

Kipp co-founded Revolution By Design (*http://www.revolutionxdesign.org*), a nonprofit education and research organization dedicated to empowerment through technology and co-organizes Rhode Island's mini Maker Faire (*http://makerfaireri.com*). As the Senior Design Engineer and Lecturer (*http://research.brown.edu/research/profile.php?id=1296749291*) at the Brown University School of Engineering, Kipp teaches several engineering design and entrepreneurship courses. He serves on the boards of The Rhode Island Museum of Science and Art (*http://www.rimosa.org*), The Providence Athenaeum (*http://www.providenceathenaeum.org*), and the community arts organization AS220 (*http://as220.org*). He is also on the technical advisory board of MAKE Magazine, is a Fellow (*http://www.philau.edu/designengineeringandcommerce/decfellows.html*) at the College of Design, Engineering and Commerce at Philadelphia University, and is an Adjunct Critic at the Rhode Island School of Design (*http://www.risd.edu/Graduate_Studies/Kipp_Bradford/*).

Colophon

The animal on the cover of *Distributed Network Data* is a guillemot (also known as a bridled common murre).

The cover image is from Cassell's *Natural History*. The cover font is Adobe ITC Garamond. The text font is Adobe Minion Pro; the heading font is Adobe Myriad Condensed; and the code font is Dalton Maag's Ubuntu Mono.

Have it your way.

O'Reilly eBooks

- Lifetime access to the book when you buy through oreilly.com
- Provided in up to four DRM-free file formats, for use on the devices of your choice: PDF, .epub, Kindle-compatible .mobi, and Android .apk
- Fully searchable, with copy-and-paste and print functionality
- Alerts when files are updated with corrections and additions

oreilly.com/ebooks/

Safari Books Online

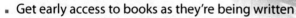

- Access the contents and quickly search over 7000 books on technology, business, and certification guides
- Learn from expert video tutorials, and explore thousands of hours of video on technology and design topics
- Download whole books or chapters in PDF format, at no extra cost, to print or read on the go
- Get early access to books as they're being written
- Interact directly with authors of upcoming books
- Save up to 35% on O'Reilly print books

See the complete Safari Library at safari.oreilly.com

O'REILLY®

Spreading the knowledge of innovators.   oreilly.com

©2011 O'Reilly Media, Inc. O'Reilly logo is a registered trademark of O'Reilly Media, Inc. 00000

Lightning Source UK Ltd.
Milton Keynes UK
UKOW021252140313

207609UK00003B/45/P